HOUSEKEEPING MANAGEMENT

Housekeeping Management

Editor

Delphina Fonseca

scitus
academics

Housekeeping Management
Edited by **Delphina Fonseca**

ISBN: 978-1-68117-546-1
Library of Congress Control Number: 2016942950

SCITUS Academics LLC,
www.scitusacademics.com
Box No. 4766, 616 Corporate Way,
Suite 2, Valley Cottage,
NY 10989

Notice

Reasonable efforts have been made to publish reliable data and views articulated in the chapters are those of the individual contributors, and not necessarily those of the editors or publishers. Editors or publishers are not responsible for the accuracy of the information in the published chapters or consequences of their use. The publisher believes no responsibility for any damage or grievance to the persons or property arising out of the use of any materials, instructions, methods or thoughts in the book. The editors and the publisher have attempted to trace the copyright holders of all material reproduced in this publication and apologize to copyright holders if permission has not been obtained. If any copyright holder has not been acknowledged, please write to us so we may rectify.

Printed in United States of America on Acid Free Paper

PREFACE

Housekeeping refers to the management of household duties and tasks such as cleaning, shopping, housekeeping, cooking, laundry and bill pay. These jobs could be performed by any of the members of the household or by various other persons hired to perform these duties. The primary duty of housekeeping is usually to protect the cleanliness and orderliness of the hotel rooms and to determine that the supplies to which guests are entitled are adequately filled. A housekeeping manager not only supervises housekeepers in the performance of their duties, but also examines some complaints that guests make and remedy. Managers supervise household employees in their duties and ensure that they meet the prescribed hotel standards.

Housekeeping Management explores the role of the housekeeping department in hotel/accommodation operations and mainly focuses on real communication between housekeeping, front office and engineering and maintenance personnel. Housekeeping duties include cleaning and turning off rooms, reserving rooms once guests leave, transporting room service trays from the hallways to the landings and verifying that all public places in the hotel are clean and orderly. The manager prepares the daily work schedule and assigns household workers to various tasks on the basis of the day's requirements. A hotel manager is responsible for encouraging team spirit among housekeepers and directing them to guest satisfaction.

TABLE OF CONTENTS

1

A DECISION PROBLEM OF HOTELS ON SETTING ONLINE-EXCLUSIVE-ROOMS IN E-COMMERCE AGE

Liang Xu, Xiaoyan Xu

University of Science and Technology of China, Hefei, China

ABSTRACT

Due to the rapid development of information technology, customers' consumption pattern in hotel industry has dramatically changed, and many people book rooms through some professional online platforms. Hotel managers need to redesign their sale channel structures by incorporating online platforms, in order to increase profit. This paper focuses on a revenue maximization problem for a hotel which cooperates with an online platform. Two cooperating forms are investigated and compared. Specifically, one intuitive form is named complete sharing, in which all the hotel rooms are shared with the online platform for booking. The other form is named setting online-exclusive-rooms (OERs), in which the hotel set a fixed quantity of rooms for exclusively booking from the online platform. By building a loss queueing model, this paper finds some properties of the optimal quantity of OERs in the latter form. Factors affecting the selection and how they affect the selection are also investigated.

KEYWORDS

Revenue Management, Hotel Room Management, Online Platform, Optimization

1. INTRODUCTION

During the last two decades, information technology has been developing rapidly. More and more people buy products or book service through the internet [1]. A typical example is that many customers prefer to booking hotel rooms via online platforms (e.g. Expedia, CTRIP, etc.), rather than through the front desk of the hotel. Customers' consumption pattern has changed and is still changing hotel industry. According to CTRIP's survey in 2010, more than 40% of hotel customers book rooms through such online platform in China, and more than 1 million rooms are booked through CTRIP per year [2]. Professional hotel booking platforms nowadays have become an important channel for hotel room booking [3] [4].

With this emerging phenomenon, hotel managers have to redesign their sale channel structures by incorporating online platforms, in order to reach different market segments, expand market share, and increase profit. One intuitive cooperating form between a hotel and an online platform is that, the hotel completely shares all rooms with the online platform for room booking. The online platform gets a commission from the hotel once it successfully accomplishes a room booking. We name this cooperating form as "complete sharing". This form seems to be reasonable and can maximize the utilization of the hotel rooms. However, this cooperating form may lead to a foreseeable result for the hotel: more and more customers will choose to book rooms via the online platform, and then the hotel should strongly rely on the online platform for getting customers. In the long term, the profit of the hotel cannot be maximized. Actually, the cooperation between CTRIP and many hotels is in another form. The hotel offers a fixed quantity of rooms to the online platform (i.e.

CTRIP) which is less than the total quantity of rooms the hotel has. This fixed quantity of rooms can only be booked from the online platform, so that we call them Online-Exclusive-Rooms (OERs). No more room can be booked from the online platform once all OERs have been booked out. We name this cooperating form as "setting OERs".

In this paper, we focus on a profit maximization decision problem for a hotel when cooperating with an online platform. Specifically, we intend to answer the question that should a hotel set OERs, and if yes, how many OERs should be set. The rest of this paper is structured as follows. In Section 2, we review the related literature and show the difference between our work and the extant research, where the contributions of this research are addressed. In Section 3, the model is formally introduced, and some properties of the optimal quantity of OERs are discussed. In Section 4, we formally discuss how parameters affect the optimal quantity of OERs and the selection of cooperating forms, by analytical study and numerical study. We conclude this paper and give some future research directions in Section 5. All the proofs of propositions are relegated to the Appendix.

2. LITERATURE REVIEW

Our work is relevant to the research on hotel yield management and the dual-channel supply chain management. However, there are some significant differences between this work and the extant literature. Next we will review the literature and the differences accordingly.

2.1. Hotel Yield Management

Much of the early literature in hotel yield management focuses on the overbooking problem, i.e. to set appropriate overbooking limits, e.g., see Rothstein [5], Liberman and Yechialli [6]. Later on, the literature on yield management focuses on maximizing the profit generated by a given capacity for service in the long run [7] [8]. In hotel industry, the capacity is the number of rooms the hotel has and the operational objective in hotel yield management is to sell the right rooms to the right customers at the right price. Bitran and Mondschein study optimal strategies for renting hotel rooms when there is a stochastic and dynamic arrival of customers from different market segments [9]. They formulate the problem as a stochastic and dynamic programming model and characterize the optimal policies as functions of the capacity and the time left until the end of the planning horizon. Bitran and Gilbert formulate a room allocation problem as a stochastic dynamic program and develop simple heuristic procedures for accepting reservations [10]. Badinelli also presents a dynamic programming formulation of the hotel yield management problem allowing for general demand patterns and a policy that is based on time and the number of vacancies [8]. Baker and Collier investigate a new Price Setting Method (PSM) to manage demand in the hotel yield management [11].

The main difference between the classical hotel yield management research and our work is that, the capacity (rooms) is usually regarded as a given resource constraint in classical hotel yield management research but a decision variable in our problem. Accompanied by the coming of E-commerce age, the problem of designing an optimal cooperating mechanism for hotel room booking service has emerged, but not investigated yet in the literature. In addition, an operational difficulty in hotel yield management is the customer's multiple night stays [9]. Generally, this problem is approximated by a series of independent single night reservations in extant hotel yield management literatures [10]. In our work, this difficulty does not exist, because the customers' stay time can be any number of days.

2.2. Dual-Channel Supply Chain Management

Most of the traditional dual-channel supply chain management research focuses on selling tangible product such as computer, which can be stored for selling at different times. The rapid development of E-commerce has made it easier for tangible product manufacturers to move part of sales online. About ten years ago, about 42% of top-ranked suppliers in a variety of industries (e.g., IBM, Pioneer Electronics, Cisco System, Estee Lauder, and Nike) have begun to sell products directly to consumers over the Internet [12]. Manufacturers nowadays are increasingly adopting a dual-channel way to sell their products, i.e., a traditional retail channel and an online channel.

Many problems regarding Dual-channel supply chain management have been extensively investigated in the literature, including the pricing and ordering decision problems in each channel, the coordination and competition problems between the dual channels [13] -[15]. Different from the extant research on traditional dual-channel supply chain management, this work considers the allocation problem of sellable "product" (the room booking service) and the corresponding profit maximizing problem.

In summary, up to now, little is known about how the hotel should use their limited room resource effectively to cooperate with the online platform. This is the main motivation for our work. The main contributions of our paper are three-folds. 1) We study the hotel room allocation problem considering the cooperation between hotel and the online platform. 2) We creatively formulate the loss-queueing model for the hotel service system which consists of a hotel and an online platform. 3) We investigate whether a hotel should set OERs and the property of the optimal quantity of OERs to maximize the hotel's revenue, which provides a theoretical guide for the management of hotel rooms in E-commerce age.

3. MODEL FORMULATION AND ANALYSIS

3.1. Problem Description

Consider a hotel service system which is composed of a hotel and an online platform. The hotel provides room service for customers, while any customer can book the room service from either the front desk or the online platform.

Here we only describe the decision problem of the hotel in "setting OERs" case. Take CTRIP as an example. As an online platform, CTRIP provides room booking service to customers for many hotels. Each hotel sets a certain quantity of rooms for booking exclusively from CTRIP, and the remaining quantity of rooms for booking exclusively from the front desk. As addressed in the first section, the former type of rooms are named "Online-Exclusive-Rooms" or "OERs", and the latter "hotel exclusive rooms" or "HERs". Once CTRIP successfully accomplishes a booking task to one customer, the hotel pays a commission to CTRIP. If CTRIP did not "book out" all its OERs, it does not have any penalty. If CTRIP has "booked out" all its OERs, no extra customer can book a room from CTRIP. Also, if the hotel has "booked out" all its HERs, it cannot accept more customers from the front desk. From this way, it will be preferable for CTRIP to have more OERs, because more OERs bring more opportunities to make money. However, the hotel needs to think about an appropriate quantity of OERs. If the quantity of OERs is too large, the quantity of HERs will be less, which leads to potential loss of the hotel's profit. If the quantity of OERs is too few, it will also lead to potential loss of customers and profit from online platform.

In this paper, we focus on the decision problem of a hotel which cooperates with an online platform. Specifically, we intend to answer the following two questions. First, within the above mentioned two cooperating forms (complete sharing and setting OERs), which one is better and when? In other words, when should a hotel set OERs? Second, in the latter form, how many OERs should the hotel set to maximize its profit? This is actually an optimization problem of the hotel. Since the second question is only related to the second cooperating form (setting OERs), we set the "complete sharing" case as benchmark, and mainly discuss the "setting OERs" case.

For simplicity and convenience of discussion, we make the following assumptions in this work.

1) All the hotel rooms are homogeneous, or of a single type. Actually, hotels usually have multiple types of rooms. For these hotels, different type of rooms can be analyzed separately.

2) Each customer only books one room at one time. This assumption is feasible, since it is the number of rooms booked that affects the revenue of the hotel, not the number of customers who make the booking.

3) There is no refund for the booking. Making this assumption is for simplicity.

4) The online platform does not set the lower threshold of OER quantity, since it will make money even if there is only one OER. In actual case, there may be a lower threshold set by the online platform. However, this constraint can be easily added to the hotel's optimal OER quantity to give the final solution.

3.2. Model for "Setting OERs"

The hotel service system investigated in this paper can be modeled by a queuing service system with multiple servers. The rooms (including OERs and HERs) can be regarded as the servers, and the whole process that customers stay in the hotel rooms can be regarded as the serve process. In our problem, there are two customer flows, one arriving at the hotel's front desk and the other arriving at the online platform. We assume that these two customer flows are independent, and each of them arrives according to a Poisson processes [5] [9] [10]. Assume that the time (days) a customer stay in the hotel (either in an OER or an HER) is an independent and identically distributed (i.i.d.) random variable obeying a general distribution, and independent of the arrival process [16] [17]. Here are the notations used in this paper.

Q: the total quantity of rooms in the hotel;

q : the total quantity of OERs;

w_e: the price when a customer books a room from the online platform;

w_h : the price when a customer books a room from the front desk;

r: the commission the hotel pays to the online platform for successfully booking out a OER; we suppose that the commission is a proportion of the booking price from the online platform;

θ: the proportion of the booking price as the commission $(0 < \theta < 1)$; thus $r = \theta \cdot w_e$;

λ_h: customers' arriving rate at the front desk (obey a Poisson process); independent of the booking prices;

λ_e: customers' arriving rate at the online platform (obey a Poisson process); independent of the booking prices;

u : service rate in the hotel (obey a general distribution); thus $1/\mu$ is the expected customer stay time (days) in the hotel.

Following the assumptions and notations above, there are $Q-q$ rooms (HERs) that can be booked from the front desk, and q rooms (OERs) from the online platform. We also note that new arriving customers will leave immediately when all the rooms (either OERs or HERs) are occupied (i.e., all servers are busy). Therefore, the hotel front desk can be regarded as an $M/G/Q-q/Q-q$ queuing system [18], and the online platform can be regarded as an $M/G/q/q$ queuing system. These two systems are independent. We illustrate them in Figure 1.

To answer the two questions raised in Section 3.1, first we need to compute the revenue of the hotel under any quantity of OERs. According to the classical queueing theory [19], the revenue of the service system depends on the expected number of customers in the service system, which in turn depends on the probabilities of all rooms taken in our problem.

Let $P_h(q)$ and $P_e(q)$ be the probabilities of all rooms taken at front desk and at online platform, respectively, when the quantity of OERs is q. Actually, $P_h(q)$ and $P_e(q)$ also represent the customer loss probability in the $M/G/Q-q/Q-q$ and $M/G/q/q$ systems, respectively. Thus $P_h(q)$ and $P_e(q)$ can be easily computed as follows [20].

$$P_h(q) = \frac{\lambda_h^{Q-q}}{\mu^{Q-q}(Q-q)!} \cdot \frac{1}{\sum_{k=0}^{Q-q} \frac{\lambda_h^k}{\mu^k k!}}, \tag{1}$$

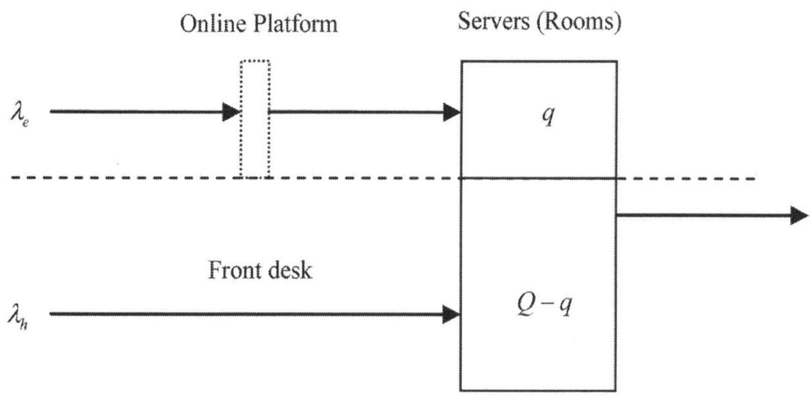

Figure 1. The front desk and online platform queuing service systems.

$$P_e(q) = \frac{\lambda_e^q}{\mu^q q!} \cdot \frac{1}{\sum_{k=0}^{q} \frac{\lambda_e^k}{\mu^k k!}} \quad (2)$$

Let $L_h(q)$ and $L_e(q)$ be the expected number of customers in the $M/G/Q-q/Q-q$ and $M/G/q/q$ system, respectively, when the quantity of OERs is q. Then $L_h(q)$ and $L_e(q)$ can be calculated as follows [21].

$$L_h(q) = \frac{\lambda_h}{\mu}\left[1 - P_h(q)\right], \quad (3)$$

$$L_e(q) = \frac{\lambda_e}{\mu}\left[1 - P_e(q)\right]. \quad (4)$$

The revenue of the hotel comes from two parts: one from the $M/G/Q-q/Q-q$ system and the other from $M/G/q/q$ system (removing the commission paid to the online platform). The expected revenue of the hotel, $\pi_h(q)$, is as follows.

$$\pi_h(q) = w_h L_h + (1-\theta) w_e L_e = \frac{w_h \lambda_h}{\mu}\left[1 - P_h(q)\right] + \frac{(1-\theta) w_e \lambda_e}{\mu}\left[1 - P_e(q)\right]. \quad (5)$$

By substituting (1) and (2) into the Equation (5), we get

$$\pi_h(q) = \frac{w_h \lambda_h}{\mu}\left[1 - \frac{\lambda_h^{Q-q}}{\mu^{Q-q}(Q-q)!} \frac{1}{\sum_{k=0}^{Q-q} \frac{\lambda_h^k}{\mu^k k!}}\right] + \frac{(1-\theta) w_e \lambda_e}{\mu}\left(1 - \frac{\lambda_e^q}{\mu^q q!} \frac{1}{\sum_{k=0}^{q} \frac{\lambda_e^k}{\mu^k k!}}\right). \quad (6)$$

Since the cost of providing room service can be regarded as a constant, it does not affect the optimal decision of the hotel. In other words, revenue maximization has the same meaning of profit maximization. In the latter part of this paper, we use revenue maximization as the operational objective of the hotel.

Before searching for the optimal decision, we need to understand the property of the expected revenue of the hotel. We proved that the expected revenue of the hotel has the following property, as shown in Proposition 1. All the proofs of propositions are arranged in the Appendix part.

Proposition 1. The expected revenue of the hotel, $\pi_h(q)$, is a concave function of the quantity of OERs, q.

Proposition 1 implies that, over any interval of the quantity of OERs (q), there is a maximal value of the expected revenue of the hotel (i.e. $\pi_h(q)$). The optimal quantity of OERs corresponding to the maximal value of $\pi_h(q)$ is what we are searching for.

3.3. The Optimal Quantity of OERs

Based on Proposition 1, we can formulate the following model to find the optimal quantity of OERs to maximize the hotel's revenue.

$$\max \pi_h(q) \quad s.t. \quad 0 \le q \le Q, \ q \in \boldsymbol{Z} \tag{7}$$

Because q is discrete, it is not easy to directly solve the above model. Denote the solution of Model (7) (i.e. the optimal quantity of OERs) to be q^*. According to Proposition 1, $\pi_h(q)$ is a concave function of q, therefore we can use the marginal analysis method to find q^* [22]. In Model (7), q^* must satisfy the following conditions:

$$\begin{cases} \pi_h\left(q^*\right) \geq \pi_h\left(q^*-1\right) \\ \pi_h\left(q^*\right) \geq \pi_h\left(q^*+1\right), \end{cases} \tag{8}$$

where $\pi_h\left(q^*\right)$ is the expected revenue of the hotel when the quantity of OERs is q^*. Based on the conditions in (8), we have Proposition 2 for determining q^*.

Proposition 2. The optimal quantity of OERs $\left(q^*\right)$ can be determined by the following conditions

$$\frac{P_h\left(q^*\right)-P_h\left(q^*-1\right)}{P_e\left(q^*-1\right)-P_e\left(q^*\right)} \leq \frac{(1-\theta)w_e\lambda_e}{w_h\lambda_h} \leq \frac{P_h\left(q^*+1\right)-P_h\left(q^*\right)}{P_e\left(q^*\right)-P_e\left(q^*+1\right)}. \tag{9}$$

The conditions in Proposition 2 are not intuitive, but the implications are reasonable and practical. Let $g(q)=\frac{P_h(q)-P_h(q-1)}{P_e(q-1)-P_e(q)}$, then $g(q+1)=\frac{P_h(q+1)-P_h(q)}{P_e(q)-P_e(q+1)}$. $g(q)$ means the ratio of the increase of customer loss probability at front desk to the decrease of customer loss probability at online platform, if the quantity of OERs is increased from $q-1$ to q. In (9), $\frac{(1-\theta)w_e\lambda_e}{w_h\lambda_h}$, which is equal to $\frac{(1-\theta)w_e\lambda_e/\mu}{w_h\lambda_h/\mu}$, means the ratio of the expected revenue per day from the online platform to that from the front desk with no customer loss. Because $P_e(q)$ and $P_h(q)$ are decreasing and increasing in q, respectively [23], and $P_e(q)$ and $P_h(q)$ are convex functions of q (see the proof of Proposition 1), then $g(q)$ is increasing in q. The left inequality in (9), $g\left(q^*\right) \leq \frac{(1-\theta)w_e\lambda_e}{w_h\lambda_h}$, is equivalent to $\left[P_h\left(q^*\right)-P_h\left(q^*-1\right)\right]w_h\lambda_h \leq \left[P_e\left(q^*-1\right)-P_e\left(q^*\right)\right](1-\theta)w_e\lambda_e$. It implies that the increase of expected revenue loss per day from the front desk is less than the decrease of expected revenue loss from the online platform, if the quantity of OERs is increased from $q^* - 1$ to q^*. The right inequality in (9), $\frac{(1-\theta)w_e\lambda_e}{w_h\lambda_h} \le g(q^*+1)$, is equivalent to $\left[P_h(q^*+1) - P_h(q^*)\right]w_h\lambda_h \ge \left[P_e(q^*) - P_e(q^*+1)\right](1-\theta)w_e\lambda_e$.

It implies that the increase of expected revenue loss per day from the front desk is larger than the decrease of expected revenue loss from the online platform, if the quantity of OERs is increased from q^* to q^*+1. Since $g(q)$ is increasing in q, the two inequalities in (9) ensure that q^* is the optimal quantity of OERs which maximizes the expected revenue of the hotel.

Up till now we are able to answer the second question proposed in Section 3.1. However, to answer the first question, we need to investigate the "complete sharing" case, and do some analysis on the property of the optimal solution of Model (7).

3.4. Model for "Complete Sharing"

In the "complete sharing" case, the available room in the hotel will be provided to the customer whose booking request comes first, either from the front desk or from the online platform. To the hotel, these two customer flows can be integrated to one new customer flow. Based on the assumptions that these two customer flows are independent Poisson arrival process, with the arrival rate of λ_h and λ_e, respectively, the new customer arrival process is also a Poisson process and the arrival rate is $\Lambda(=\lambda_h+\lambda_e)$ [24]. Since it is assumed that the time customers stay in the hotel is an i.i.d. random variable and follows a general distribution, the front desk together with the online platform is an $M/G/Q/Q$ queue system.

Let $\tilde{P}$ be the probability of all rooms taken in the $M/G/Q/Q$ system, then

$$\tilde{P} = \frac{(\lambda_h + \lambda_e)^Q}{\mu^Q Q!} \frac{1}{\sum_{k=0}^{Q} \frac{(\lambda_h + \lambda_e)^k}{\mu^k k!}} . \tag{10}$$

For the convenience of comparison, we let the booking prices and the commission proportion remain the same for both the "setting OERs" case and the "complete sharing" case. Let $\tilde{\pi}$ be the expected revenue of the hotel in the "complete sharing" case. $\tilde{\pi}$ can be computed as follows.

$$\tilde{\pi} = \frac{w_h \lambda_h + (1-\theta) w_e \lambda_e}{\mu} (1 - \tilde{P}) = \frac{w_h \lambda_h + (1-\theta) w_e \lambda_e}{\mu} \left(1 - \frac{(\lambda_h + \lambda_e)^Q}{\mu^Q Q!} \frac{1}{\sum_{k=0}^{Q} \frac{(\lambda_h + \lambda_e)^k}{\mu^k k!}} \right). \tag{11}$$

It is obvious from Equation (11) that $\tilde{\pi}$ has no relationship with q. That's why we set the "complete sharing" case as the benchmark.

4. PROPERTY ANALYSIS

To find out when should a hotel set OERs and when a hotel should share all room resources with the online platform, we check the relative magnitude of the expected revenue of the hotel under "setting OERs" and "complete sharing" cases. Define a variable α to be the ratio of these two expected revenues, i.e.,

$$\alpha = \frac{\pi_h (q^*)}{\tilde{\pi}} . \tag{12}$$

when $\alpha > 1$, the hotel is more willing to set OERs. And when $\alpha < 1$, the hotel is more willing to share all room resources with the online platform.

Since $\pi_h(q^*)$ depends on q^*, α also depends on q^*. Thus in the remaining part of this section, we try to finish two tasks. One is to find out what variables affect the optimal quantity of OERs, q^*, and what the effects are. The other is to find out what variables affect the revenue ratio, α, and what the effects are.

Specifically, we check the effects of the following four factors on q^* and α: the booking prices (w_h and w_e), the proportion of the booking price as the commission(θ), the customer arriving rates (λ_h and λ_e), and the expected customer stay time in the hotel $(1/\mu)$.

4.1. The Factor of Booking Prices (w_h and w_e)

For the effects of booking prices (w_h and w_e) on q^*, we have the following proposition.

Proposition 3: The optimal quantity of OERs, decreases in w_h, and increases in w_e.

Proposition 3 can be explained as follows. When the other parameters are given, if w_h is increased, the potential revenue from a room booked at the hotel's front desk is increased. Thus the hotel is more willing to limit rooms for booking at the online platform, i.e. setting less OERs. However, if w_e is increased, the revenue from a room booked at the online platform front desk is increased. Thus the hotel is willing to offer more OERs to the online platform.

Proposition 3 addresses how q^* will be changed when one booking price (w_h or w_e) is changed given the other fixed. However, when the two prices are changed simultaneously, it is difficult to find conclusions analytically. Thus we use numerical studies to analyze it.

Let $\lambda_h = 30$, $\lambda_e = 20$, $Q = 100$, $\mu = 0.5$ and $\theta = 0.2$. The effects of w_h and w_e on q^* are illustrated in Figure 2.

Figure 2 shows that the conclusions in Proposition 3 are valid. Furthermore, it can be observed fromFigure 2 that, if the ratio of w_h to w_e is increased, q^* tends to be lower. If the ratio of w_h to w_e is decreased, q^* tends to be higher. These two observations are reasonable, because higher ratio of w_h to w_e means less revenue from the online platform. Thus the hotel tends to decrease OERs.

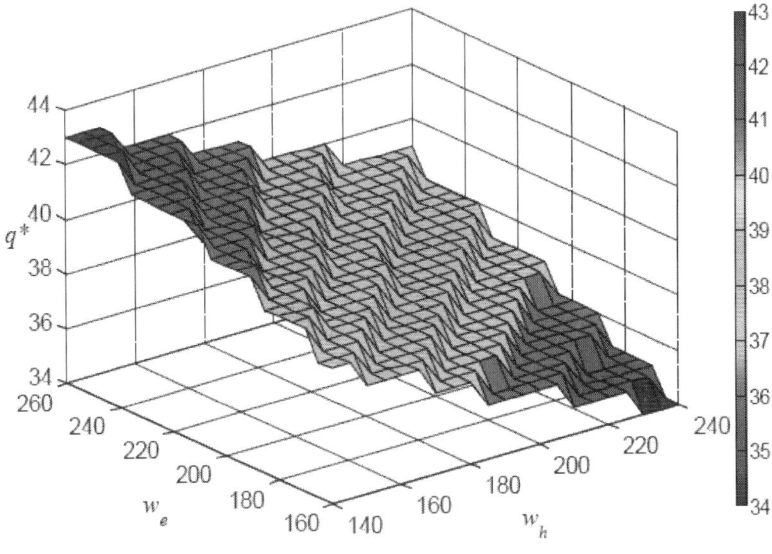

Figure 2. The effect of w_h and w_e on q^*.

It is rather difficult to analytically identify the effects of w_h and w_e on α, since $\pi_h(q^*)$ is difficult to be analytically solved. We use numerical studies to check the effects.

Let $\lambda_h = 30$, $\lambda_e = 20$, $Q = 100$, $\mu = 0.5$ and $\theta = 0.2$. The effects of w_h and w_e on α are illustrated in Figure 3.

It is shown in Figure 3 that α increases in w_h, and decreases in w_e. Furthermore, α increases in the ratio of w_h to w_e. These observations are reasonable, because on one hand, as w_e increases, the hotel can get more revenue from booking out one room at the online platform. Thus the hotel is more willing to share more rooms with the online platform, and then the "complete sharing" case is welcome. As w_e decreases, the hotel is more willing to set OERs. It is important to note that the expected revenue of the hotel in "setting OERs" case can be higher than that in "complete sharing" case, when the ratio of w_h to w_e is rather large. This indicates that in particular case, "complete sharing" form may not be better than "setting OERs" form for the hotel.

4.2. The Factor of the Proportion of Booking Price as Commission (θ)

For the effect of the proportion of booking price as commission (θ) on q^*, we have the following proposition.

Proposition 4. The optimal quantity of OERs, q^*, decreases in θ.

The conclusion in Proposition 4 can be interpreted as follows. Larger θ means the hotel should give more commission to the online platform, which implies less revenue from booking out one room at the online platform. Thus the hotel tends to set less OERs when θ is increased.

To show the effect of θ on q^* numerically, we also do some computation. Let $\lambda_h = 30$, $\lambda_e = 20$, $w_h = 200$, $w_e = 180$, $Q = 100$, and $\mu = 0.5$.

The effect of the proportion of the booking price as the commission, θ, on q^* is illustrated in Figure 4.

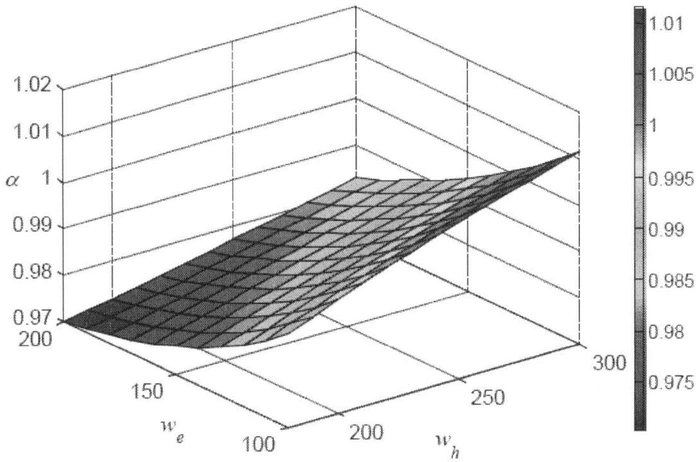

Figure 3. The effect of w_h and w_e on α.

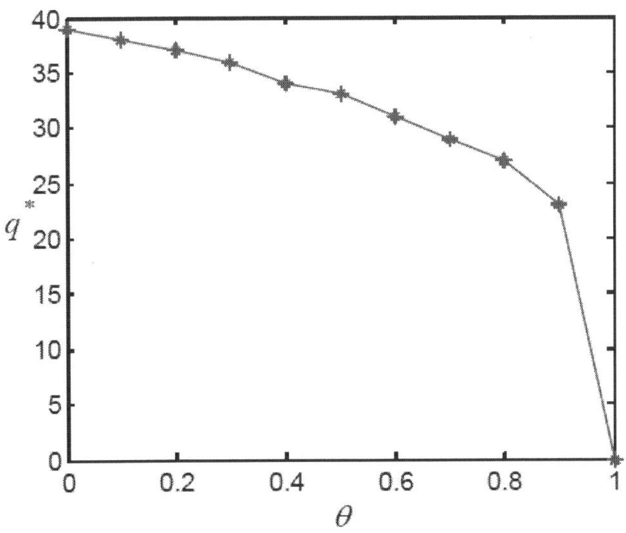

Figure 4. The effect of θ on q^*.

Figure 4 shows that q^* decreases in θ, which verifies the conclusion in Proposition 4. Let $\lambda_h = 30$, $\lambda_e = 20$, $w_h = 200$, $w_e = 180$, $Q = 100$, and $\mu = 0.5$. We check the effect of θ on α, as shown in Figure 5.

Figure 5 shows that α increases in θ, which means as θ increases, the hotel is less willing to share all room resources with the online platform but like to set OERs. This is reasonable because higher θ means less revenue from booking out one room at the online platform.

4.3. The Factor of Customer Arriving Rate (λ_h and λ_e)

It is difficult to analytically check the effect of λ_h and λ_e on q^* and α. We use numerical studies to check these effects. Let $w_h = 200$, $w_e = 180$, $Q = 100$, $\mu = 0.5$ and $\theta = 0.2$. The effects of λ_h and λ_e on q^* are illustrated in Figure 6.

Figure 6 shows that q^* decreases in λ_h and increases in λ_e. Furthermore, q^* decreases in the ratio of λ_h to λ_e. This is because larger λ_e means more customers arriving at online platform, then it will be better to share more rooms with the online platform. However, larger λ_h means more customers arriving at the front desk. Since the total rooms are fixed, it will be better to set smaller OERs.

For the effect of λ_h and λ_e on α, it is not easy to make a speculation. We also resort to numerical studies. Let $w_h = 200$, $w_e = 180$, $Q = 100$, $\mu = 0.5$ and $\theta = 0.2$. The effects of λ_h and λ_e on α are illustrated in Figure 7.

As shown in Figure 7, the effects of λ_h and λ_e on α are a little complicated. When both λ_h and λ_e are low, α is close to 1. It means the expected revenue of the hotel is nearly the same in both the "setting OERs" case and the "complete sharing" case, because there is little probability that all rooms are taken. When one of λ_h and λ_e is high and the other is low, α is less than 1. This is because in the "complete sharing" case, all room resources can be fully utilized. But the resource utilization may not be so high in the "setting OERs" case. Thus the "complete sharing" case is better than the "setting OERs" case. If both λ_h and λ_e is high, α is higher than 1. This is because in this situation, the hotel can get highest revenue by setting a small quantity of OERs. However, if the hotel shares all room resources with online platform, many rooms may be occupied by online customer and the hotel will have potential revenue loss.

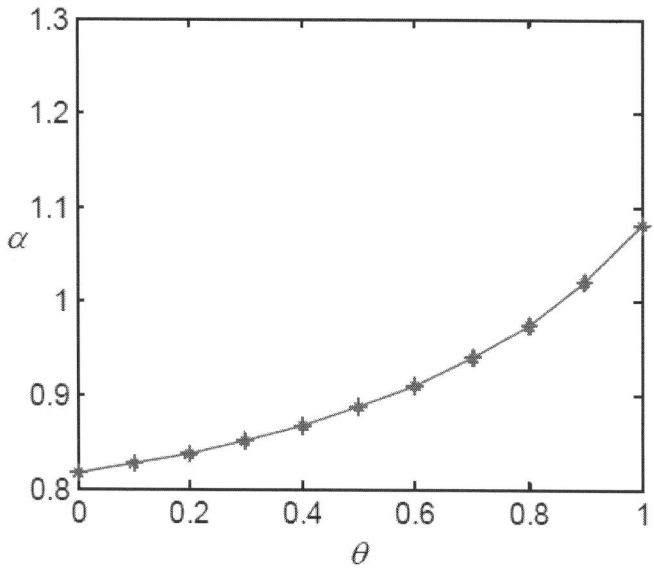

Figure 5. The effect of θ on α.

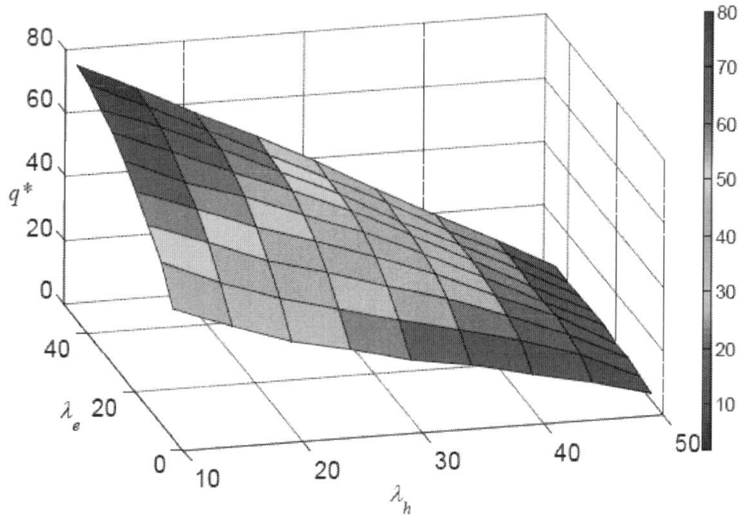

Figure 6. The effects of λ_h and λ_e on q^*.

4.4. The Factor of Expected Customer Stay Time in the Hotel $(1/\mu)$

We resort to numerical studies to check the effects of $1/\mu$ on q^* and α. Let $\lambda_h = 40$, $\lambda_e = 30$, $w_h = 200$, $w_e = 180$, $Q = 100$ and $\theta = 0.2$. The effect of the expected customer stay time $(1/\mu)$ on q^* is illustrated in Figure 8.

It can be observed from Figure 8 that q^* decreases in the expected customer stay time, $1/\mu$. This is because, as the expected customer stay time increases, the probability of all room taken in front desk is increasing. It is better to set more HERs and less OERs. When the expected customer stay time is rather long (3 or more days in this example), the optimal quantity of OERs is 0, since the probability of all room taken in front desk is close to 1, and there is no need to provide OERs to online platform. This is actually the third case, i.e. the "operating alone" case. In this case, the hotel operates alone, without cooperating with the online platform. This case is a special one of the "setting OERs" case.

The corresponding effect of $1/\mu$ on α is illustrated in Figure 9.

It can be observed from Figure 9 that α is increasing in the expected customer stay time, $1/\mu$. This can be explained similarly. As the expected customer stay time increases, the probability of all room taken in front desk is increasing. It is better to share less room with the online platform, i.e. "complete sharing" is not welcome to the hotel.

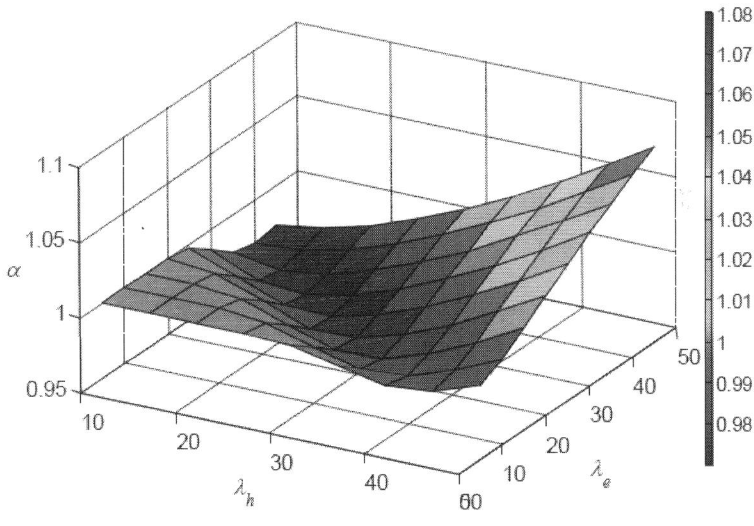

Figure 7. The effects of λ_h and λ_e on α.

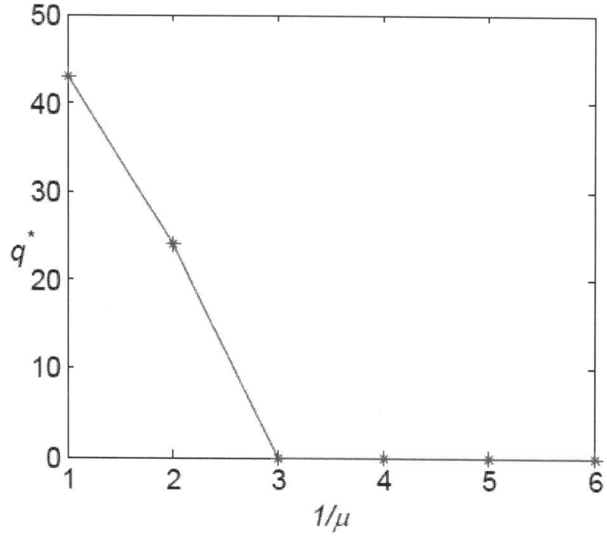

Figure 8. The effect of $1/\mu$ on q^*.

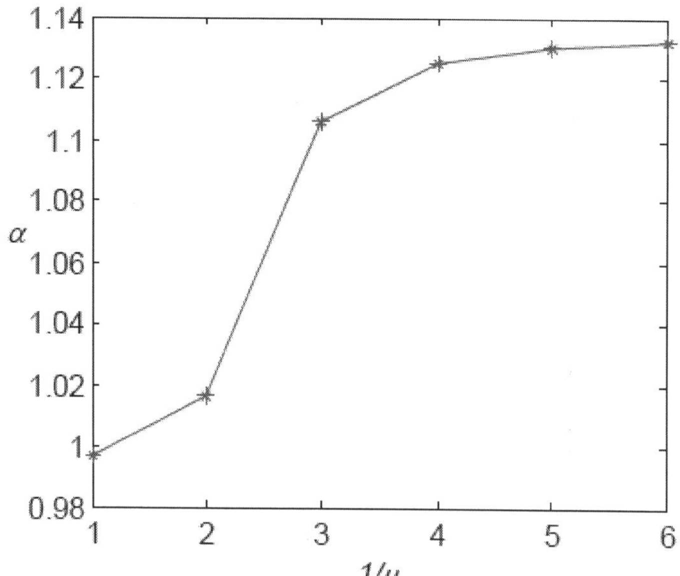

Figure 9. The effect of $1/\mu$ on α.

5. CONCLUSIONS

In this paper, we have investigated a new type of hotel room management problem in the coming E-commerce age: deciding whether to set a fixed quantity of Online-Exclusive-Rooms (OERs) for booking at the online platform, and if yes, then how many OERs should be set? This "setting OER" form is compared with another intuitive cooperating form, complete sharing, i.e. the hotel shares all room resources with the online platform for booking. We formulate the hotel and the online platform as a loss queueing service system. Properties of the expected revenue of the hotel and the optimal quantity of OERs to maximize the hotel's revenue are identified.

In the "setting OER" form, the expected revenue of the hotel is proved to be a concave function of the quantity of OERs, which insures that there is an optimal quantity of OERs which maximizes the expected revenue of the hotel. The method of determining the optimal quantity of OERs is presented. It is found by analytical and numerical studies that: 1) the optimal quantity of OERs decreases in the booking price at the front desk of the hotel, and increases in the booking price at the online platform; 2) the optimal quantity of OERs decreases in the proportion of booking price as commission; 3) the optimal quantity of OERs decreases in the customer arrival rate at the front desk, and increases in the customer arrival rate at the online platform; 4) the optimal quantity of OERs decreases in the expected customer stay time. These properties can provide suggestions on how to set and change the optimal quantity of OERs to maximize the expected revenue of the hotel.

It seems that the "complete sharing" form can utilize the capacity of the hotel more fully. The numerical studies in this paper verified this point. However, it is also shown by numerical studies that completely sharing rooms with online platform is not always a good idea for the hotel. It is better to set OERs when the online booking price is too low, or the commission part is too high, or both online and offline customer arrival rates are large, or the expected customer stay time is long.

There are several limitations in this paper. First, the booking price of the OERs is exogenously fixed but not a decision variable; we do not consider the pricing behavior of the online platform. Second, since this paper focuses on the profit optimization of the hotel, we do not consider the behaviors of the online platform, such as advertizing and ranking which can attract customers to book from the online platform. Third, we assume that no more customers can book rooms from the online platform once all OERs are booked out. Actually, some customers still can get rooms in the above case, but with a lower probability. These limitations are all interesting future research directions. Furthermore, if the hotel adopts the "complete sharing" form, how should the expected revenue of the hotel change in the long run? This should also be an interesting topic in the future.

REFERENCES

1. Dellarocas, C. (2003) The Digitization of Word-of-Mouth: Promise and Challenges of Online Feedback Mechanisms. Management Science, 49, 1407-1424.

2. China Tourism Academy (2011) Report of the Development of China's Tourism E-Commerce.

3. Morgan, N.P. and Piggott, A.R. (2003) Destination Branding and the Role of the Stakeholders: The Case of New Zealand. Journal of Vacation Marketing, 9, 285-299.

4. O'Connor, P. and Frew, J.A. (2004) An Evaluation Methodology for Hotel Electronic Channels of Distribution. International Journal of Hospitality Management, 23, 179-199.

5. Rothstein, M. (1974) Hotel Overbooking as a Markovian Sequential Decision Process. Decision Science, 5, 289-404.

6. Liberman, V. and Yechialli, U. (1978) On the Hotel Overbooking Problem: An Inventory System with Stochastic Cancellations. Management Science, 24, 1117-1126.

7. Sieburgh, J.A. (1988) Yield Management at Work at Royal Sonesta. Lodging Hospitality, 9, 235-237.

8. Badinelli, R.D. (2000) An Optimal, Dynamic Policy for Hotel Yield Management. European Journal of Operational Research, 121, 476-503.

9. Bitran, G. and Mondschein, S.V. (1995) An Application of Yield Management to the Hotel Industry Considering Multiple Stays. Operation Research, 43, 427-443.

10. Bitran, G. and Gilbert, S.M. (1996) Managing Hotel Reservations with Uncertain Arrivals. Operation Research, 44, 35-49.

11. Baker, R.D. and Collier, D.A. (2003) The Benefits of Optimizing Prices to Manage Demand in Hotel Revenue Management Systems. Production and Operations Management, 12, 502-518.

12. Chiang, W.K., Chhajed, D. and Hess, J. (2003) Direct Marketing, Indirect Profits: A Strategic Analysis of Dual-Channel Supply-Chain Design. Management Science, 49, 1-20.

13. Tsay, A. and Agrawal, N. (2004) Channel Conflict and Coordination in the E-Commerce Age. Production and Operations Management, 13, 93-110.

14. Viswanathan, S. (2005) Competing across Technology-Differentiated Channels: The Impact of Network Externalities and Switching Costs. Management Science, 51, 483-496.

15. Dumrongsiri, A., Fan, M., Jain, A. and Moinzadeh, K. (2008) A Supply Chain Model with Direct and Retail Channels. European Journal of Operational Research, 187, 691-718.

16. Miller, B.L. (1969) A Queuing Reward System with Several Customer Classes. Management Science, 16, 234-245.

17. Printezis, A., Burnetas, A. and Mohan, G. (2009) Pricing and Capacity Allocation under Asymmetric Information Using Paris Metro Pricing. International Journal of Operational Research, 5, 265-279.

18. Kimura, T. (2000) Equivalence Relations in the Approximations for the M/G/s/s+r Queue. Mathematical and Computer Modeling, 31, 215-224.

19. Medhi, J. (2003) Stochastic Models in Queueing Theory. Academic Press, Boston.

20. Hassin, R. and Haviv, M. (2003) To Queue or Not to Queue: Equilibrium Behavior in Queuing Systems. Kluwer Academic Publishers, Boston.

21. Kleinrock, L. (1975) Queueing Systems. Vol. I, Wiley, New York.

22. Fox, B. (1966) Discrete Optimization via Marginal Analysis. Management Science, 13, 210-216.

23. Harel, A. (1990) Convexity Properties of the Erlang Loss Formula. Operations Research, 38, 499-505.

24. Rabi, N.B. and Edward, C.W. (2009) Stochastic Processes with Applications. Society for Industrial and Applied Mathematics, Philadelphia.

25. Messerli, E.J. (1972) Proof of a Convexity Property of the Erlang B Formula. The Bell System Technical Journal, 51, 951-953.

26. Boyd, S. and Vandenberghe, L. (2004) Convex Optimization. Cambridge University Press, Cambridge.

2

POSTTRAINING SELF-EFFICACY, JOB INVOLVEMENT, AND TRAINING EFFECTIVENESS IN THE HOSPITALITY INDUSTRY

Xinyuan Zhao[a] & Karthik Namasivayam[b]

[a]School of Tourism and Hotel Management, South China University of Technology, Guangzhou, China

[b]School of Hospitality Management, Pennsylvania State University, University Park, PA

ABSTRACT

The current study proposed that in the context of hospitality training (1) posttraining self-efficacy mediates the effects of training acquisition on trainees' behavioral intentions to apply learning outcomes, and (2) posttraining self-efficacy either fully or partially mediates the effects of job involvement on trainees' behavioral intentions. Data were collected in a laboratory experiment conducted in the context of self-checkout machines in a university library. The results of a path analysis supported the mediation effects of posttraining self-efficacy on the relationship between training acquisition and trainees' behavioral intentions. However, the results did not support the mediation effects of posttraining self-efficacy on the relationship between job involvement and trainees' behavioral intentions. This suggests an independent model of individual and situational factors on training effectiveness in the hospitality industry. After discussing the limitations of the research, the article concludes with managerial implications for hotel managers and directions for future studies.

KEYWORDS

Training effectiveness, training motivation, posttraining self-efficacy, job involvement

1. INTRODUCTION

Training has become a huge investment in organizations. Von Bergen and Mawer (2007) report that the companies represented in *Training Magazine's* Training Top 100 invested about $6 billion and about 65 hours of training per employee; other organizations spend about $800 per employee annually. It is apparent that training costs are substantial and organizations seek evidence that the training money spent is justified (Tracey & Tews, 1995).

Both theory and practice have improved dramatically over the last few decades in demonstrating training effectiveness (Salas & Cannon-Bowers, 2001). Research has identified various factors that contribute to the effectiveness of training programs. Pretraining self-efficacy was identified as one critical cause (Tracey, Hinkin, Tannenbaum, & Mathieu, 2001). Training processes can be usefully arranged on a timeline with three distinct phases: pretraining, training, and posttraining; much previous research has focused on self-efficacy before or during training (Salas & Cannon-Bowers, 2001). Little research exists exploring the links between self-efficacy and training effectiveness in the posttraining phase.

Different than existing research, the current study focuses on the effects of posttraining self-efficacy on training effectiveness. More specifically, the objective of the current study is to investigate the influence of training acquisition and job involvement on posttraining self-efficacy and the relationship of posttraining self-efficacy and training effectiveness. It is suggested in this study that the degree to which an individual perceives him- or herself to have acquired useful knowledge in training will affect the individual's perceptions of self-efficacy. This relationship depends in part on the extent to which the individual is involved in their job. Individuals who perceive a higher level of self-efficacy consequent to training will be more likely to translate their learning into behaviors in the workplace. Hence posttraining self-efficacy is considered to be an important factor in translating training initiatives into work behaviors. The conceptual model is shown in Figure 1. The next section provides a review of extant literature and develops hypotheses. Next, the method and the results are presented. The article concludes with a discussion of the results and the study's limitations.

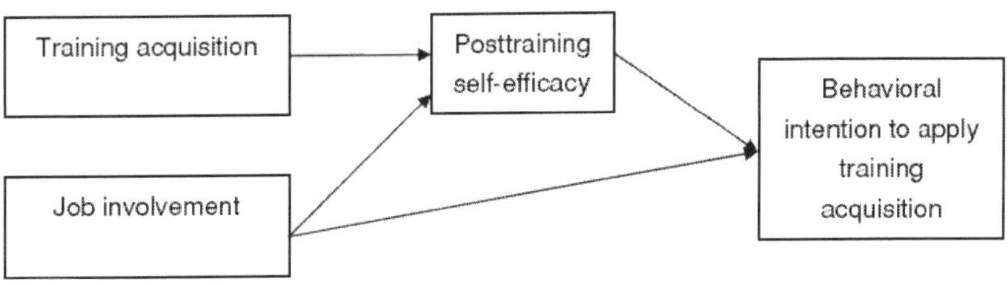

Figure 1. The conceptual model.

2. LITERATURE REVIEW

2.1 Research and Practice in Hospitality Industry Training

Employee training has become increasingly important for hotels to improve service quality, decrease labor costs, increase productivity and profitability, and effectively manage workforce diversity (Kim, 2006). For example, a highly targeted management training program helped Red Carnation reduce employee turnover rate from 80% to 30% and save 200,000 per year (Pollitt, 2006). Hyatt Regency Scottsdale worked closely with local education institutions and successfully coordinated links between corporate, local government, and educational organizations, thereby strengthening their supply of qualified and skilled labor (Enz & Siguaw, 2000). In order to get employees involved in and committed to the organization, Disney's Polynesian Resort generated three training modules, including socialization to the hospitality industry and the culture of the hotel, association between personal values and the resort, and service standards (Enz & Siguaw, 2000). However, as Dixie Budke, the general manager of Simpson House Inn in Santa Barbara, California, pointed out, only "intrinsically motivated" employees with high involvement in the job and strong self-efficacy will obtain the best outcomes from training programs (Enz & Siguaw, 2000).

Previous hospitality training research mainly focused on training needs analysis and measurement of training outcomes. Training needs in hospitality establishments are influenced by the organization's strategic goals and organizational realities (Lee, Nam, Park, & Lee, 2006). Adopting a service-oriented strategy required hotels to provide employee training on appropriate customer-oriented behaviors and reinforce factors influencing the implementation of service quality initiatives (e.g., job involvement) (Lee et al., 2006). In addition, Grau-Gumbau, Agut-Nieto, Llorens-Gumbau, and Martinez-Martinez (2002)found that new technologies, social changes, and service quality control systems are a modern hotel organization's challenges and require the greatest training investments. To further complicate the issue, training needs analysis of the skills customer-contact employees need to provide qualified services also changes across various hotel functions (e.g., housekeeping, front office, and food and beverage management) and different countries (e.g., Western and Eastern countries) (Chartrungruang, Turner, King, & Waryszak, 2005). Although the hospitality industry has expended a lot of effort and investment on employee training, the return-on-investment of hospitality training has not been satisfactory (Zhao, Zhan, & Namasivayam, 2004). As a result, training effectiveness becomes an important point of focus in hospitality training research.

2.2 Training Effectiveness

Training is "a planned effort by a company to facilitate employees' learning of job-related competencies" (Noe, 2002, p. 4). The aim of any training program is to enhance business performance. For any training program to be considered effective, trainees have to learn the training content and then apply such learning in the workplace. The effectiveness of a training program can thus be conceptualized as being composed of training acquisition and transfer of training (Tracey et al., 2001). Training acquisition refers to the learning of material or content administered in a training program by trainees. Scholars suggest that not only learning but also retention of content is important in accounting for training acquisition. A number of factors, which are outside the scope of this study, have been identified that influence training acquisition and retention. For the purposes of this article, it is suggested that once trainees have acquired learning it is important that they transfer such learning to the workplace. Thus the transfer of training is an important indicator of training effectiveness (Baldwin & Ford, 1988). In this context, the transfer of training is indicated by behavioral changes in the workplace.

There is a well-developed body of literature in training evaluation. Kirkpatrick's (1996) four-level model has been the most widely cited. In this model, training effectiveness is evaluated using four criteria: trainee reaction, learning, behavior, and business results. According to the model, training effectiveness depends on the degree to which it elicits positive affective reactions from employees immediately after training, the degree to which learning of content occurs, the extent to which the training program leads to desired changes in workplace behavior, and finally, the degree to which the training program leads to anticipated business results. Researchers have elaborated on specific levels of the model. For example, to ensure better measurement, trainees' reactions were conceptualized as consisting of utility- and affective-type reactions (Alliger, Tannenbaum, Bennett, Traver, & Shotland, 1997). Affective reactions are measured using responses to questions such as "I found this training program to be enjoyable." Utility measurements tap into the degree to which trainees feel the training program will benefit their ability to perform their jobs. Other attempts at parsing Kirkpatrick's model involved the conceptualization of a multidimensional view of learning in which changes in cognitive, affective, or skill-based outcomes were accounted for (Kraiger, Ford, & Salas, 1993). This conceptualization expanded on the concept of learning and noted that learning outcomes can be measured either as cognitive, skill-based, or affective outcomes.

Alliger et al. (1997) conducted a meta-analysis of research in training and report a large proportion of research studies considered the relation between trainee reactions and transfer of training, with only moderate correlations apparent between utility reactions and transfer. Further, only a moderate relationship was found between learning and retention and behavioral changes in the workplace in the long term. These results suggest that it is important to account for factors that may intervene between the training program and the actual application of learning in the workplace, as evidenced by behavioral changes on the job. Broadly it is argued that, given a level of learning, individuals' level of job involvement and self-efficacy will moderate their intentions to apply learning acquired during training in the workplace.

2.3 Pretraining and Posttraining Process

As noted earlier, pretraining, during training, and posttraining processes are identified as distinct stages (Salas & Cannon-Bowers, 2001). Hence the effects of various factors on training effectiveness should be estimated separately at each stage. The authors also note (p. 486) "events that occur[s] after training are as important as those that occur before and during training." Accordingly, the posttraining stage refers to the period from the end of the training program to the transfer of training, that is, the application of learning through behavior.

2.4 Posttraining Self-Efficacy

Self-efficacy is an important variable that mediates the impact of dispositional variables on learning performance (Salas & Cannon-Bowers, 2001). Self-efficacy refers to the perceived ability of an individual to engage in carrying out certain assignments to attain a specific purpose (Bandura, 1977). It is possible that individuals either have self-efficacy before going into the program or acquire it during the training (Salas & Cannon-Bowers, 2001). Haccoun & Saks (1998) suggest "training increases self-efficacy and self-efficacy mediates the effects of training on training outcomes" (p. 39). Posttraining self-efficacy performs two functions: (1) it causes individuals to reflect on their educational/learning outcomes and conclude that they are "able to do"; and (2) this knowledge has a motivational property and increases an individual's willingness to perform (willing to do). While Haccoun and Saks adopt an optimistic note, it is possible that training program designs may actually lead to a reduction in self-efficacy. This is important to note because the nature of the training program (content and learning difficulty) influences the individual's level of self-efficacy, with consequences for training transfer. If the training content facilitates the acquisition of learning, then this enhances self-efficacy levels and a positive effect on the transfer of training can be expected. Individual-level variables (cognitive ability, for example) no doubt play a role in the self-efficacy of individuals—however, it is argued in this article that self-efficacy subsumes the effects of these variables. Accordingly, posttraining self-efficacy has been defined in here as individuals' perceived ability to behave as prescribed by the training program.

2.5 Job Involvement

Job involvement is "the degree to which the job situation is central to the individual and his or her identity" (Blau, 1985, p. 34). Job involvement reflects individuals' dispositional differences about job centrality, and therefore affects training outcomes (Colquitt, LePine, & Noe, 2000). Employees with higher job involvement are more likely to acquire learning (training acquisition) valuable to their job and organization (Tracey et al., 2001) and have a higher motivation to learn. Job involvement operates on training effectiveness via two distinct pathways. The effects of job involvement on training transfer may be either fully or partially mediated by learning motivation (Colquitt et al., 2000). In the current study, posttraining self-efficacy subsumes motivation to apply training in the workplace. When training content enhances posttraining self-efficacy and individuals have high levels of job involvement, a positive relationship between job involvement and transfer of training is expected. The relationship between job involvement and transfer thus depends on the level of self-efficacy (posttraining). It is also possible, however, that individuals who express high levels of job involvement will also transfer training despite their levels of self-efficacy. Accordingly, in the full mediation framework, the effect of individuals' job involvement on transfer of training is mediated by their level of posttraining self-efficacy. The partial mediation model, on the other hand, suggests that job involvement has a direct influence on transfer of training as well as being mediated by posttraining self-efficacy.

3. METHODS

Based on a review of the literature, we developed the hypotheses stated below.

Hypothesis 1. Training acquisition positively influences posttraining self- efficacy.

Hypothesis 2. Posttraining self-efficacy positively leads to trainees' behavioral intentions to apply training acquisition.

Hypothesis 3a. Posttraining self-efficacy fully mediates the effects of job involvement on trainees' behavioral intentions to apply training acquisition.

Hypothesis 3b. Posttraining self-efficacy partially mediates the effects of job involvement on trainees' behavioral intentions to apply training acquisition.

To test the hypotheses, we conducted a laboratory experiment in the context of training students to use self-checkout machines (SST) in a university library. Thus the training focuses on service procedures and capabilities, which became a simulation of hospitality training.

3.1 Sample

Research participants were 87 college students within the hospitality management major enrolled at a large northeastern university. Participants were recruited from several courses and were offered course extra credit in return for their voluntary participation. All 87 participants took part in and completed all requirements of the research process.

3.2 Procedure

Before the study we asked the students whether they had used SSTs in the library and ensured that the participants had no prior experience in using these machines. Therefore all participants were at the same level of knowledge about using SSTs. Each participant was required to undergo the same training program explaining the operation of SSTs in the library. The participants were then asked to apply the learning and subsequently answered a self-administered paper-and-pencil survey instrument.

3.3 Training

The training programs offered by the university library on using SSTs are voluntary programs and were the focal point for the current study. The librarians who provided training were employees experienced in helping patrons use the SSTs. While recruiting the librarians to train the participants, the researchers explained to the librarians the purpose of the study and decided on the content in advance so that all participants received the same training.

In terms of the actual experimental procedure, first, a librarian handed out written instructions to the participants. The librarian then explained the instructions and answered any questions from the participants. Next, participants formed a circle and were shown step-by-step how to operate the SSTs. After participants said they were clear about their usage, each participant was asked to try out a machine by using it to check out a book. The library circulation services provided the books to be checked out.

3.4 Measures

3.4.1 Posttraining Self-Efficacy

Posttraining self-efficacy was measured in terms of the participants' expectations after training that they could successfully operate an SST (Bandura, 1977). An eight-item scale was revised for the purpose of the current study (Jones, 1986). The eight items were scored on a 7-point Likert-type scale anchored by "strongly disagree" and "strongly agree." The Appendix lists the items used in the current study.

3.4.2 Job Involvement

Job involvement evaluates the centrality and importance of the job for employees and a nine-item scale was adapted (Reeve & Smith, 2000). The modified items are presented in the Appendix.

3.4.3 Training Acquisition

Training acquisition evaluates trainees' learning outcomes. To measure training acquisition, we recruited and asked three coders to record the number of trials and time in seconds it took each subject to check out a book. Three coders were recruited from among the Honors students. We explained the procedure of recording to the coders and conducted three trials runs. When we found they had a high convergent level of recording and they felt confident to record, they took the role of recording the number of trials and the time subjects spent checking out books using the machine.

After the records were obtained, the data were coded using a special procedure. Because the number of trials and time subjects spent negatively indicate the subjects' learning, the data were reversely coded. In addition, because the number of trials and time subjects spent were close to a Poisson distribution, the data were normalized with a mean of 0 and a standardized deviation of 1 ($N(0, 1)$). Specifically, the following transformations were conducted: (1) number of trials and time in seconds for each trial for each subject were multiplied, yielding a score termed "TA1"; (2) TA1 was transformed to $N(0, 1)$ and labeled "TA2"; and (3) TA2 was centered, providing a score call "TA3." TA3 was used in data analysis as the indicator of training acquisition.

3.4.4 Behavioral Intentions

Behavioral intentions assess the subjects' intention to reuse SSTs after training and one trial. Two items developed byDabholkar and Bagozzi (2002) were modified for the library setting ("How likely is it that you will continue using the self-checkout machines" and "I intend to use the skills acquired from the training program to check out books by myself.") The items were measured on a 7-point Likert scale ranging from "likely" to "unlikely" and from "strongly disagree" to "strongly agree", respectively.

4. RESULTS

4.1 Description of the Study Variables

Table 1 describes the means, standard deviations, and Cronbach's α values of the variables in the model and provides the correlation coefficients of the variables. The results of the bivariate correlation analyses showed that except for job involvement, all correlation coefficients were significant at $\alpha = 0.05$. The remaining indicators yielded Cronbach's α values close to or above the acceptable level of 0.6 (Hair, Anderson, Tatham, & Black, 1998).

Table 1. Means, Standard Deviations, and Intercorrelations of the Study Variables

Variable	Mean	*SD*	1	2	3	4
(1) Job involvement (JI)	3.92	0.85	(0.67)			
(2) Training acquisition (TA)	5.21	1.00	−0.20	—		
(3) Posttraining self-efficacy (PTSE)	6.19	0.84	−0.14	0.31**	(0.80)	
(4) Behavioral intentions (BI)	5.45	1.47	0.19	0.23*	0.30**	(0.59)

Note. Alpha coefficients are shown in parentheses on the diagonal when applicable.
$^{*}p < .05$, $^{**}p < .01$.

4.2 Sample Size and Data Transformation

Relatively small sample sizes (in this study $N = 87$) do not preclude the use of appropriately constructed statistical analyses (e.g., Kramer & Rosenthal, 1999). The data of the study variables were standardized ($\sim N(0, 1)$) prior to analyses to ensure that the tests are unbiased. Further, effect sizes (partial η^2) were also calculated to aid in the interpretation of the results. As Kramer and Rosenthal (1999, p. 75) suggest, "effect size estimates are unaffected by the size of the study and are essential to determining the practical importance of a study." As shown in Table 2, the effect sizes in the current study range from 0.272 to 0.662, supporting the statistical and practical significance of the study.

4.3 Path Analysis

The sample ($N = 87$) was inadequate for structural equation modeling (SEM). Therefore a path analysis based on a series of regression analyses was conducted to test hypotheses (Baron & Kenny, 1986). Table 2 provides the results of the path analyses. Figure 2 illustrates the model tested in the current study. All paths in the figure are significant ($\alpha = 0.05$).

Hypothesis 1 stated that training acquisition will affect the self-efficacy of participants posttraining. The data suggest that training acquisition had a significant impact on posttraining self-efficacy ($\beta = 0.311$, $p = .003$, $\eta^2 = 0.468$). Thus Hypothesis 1 is supported.

According to Baron and Kenny (1986), mediation is indicated if (1) the independent variable significantly influences the mediator; (2) the independent variable significantly affects the dependent variable; and (3) full mediation is obtained if the independent variable has no significant impacts when the mediator is introduced in the model, while partial mediation holds if the independent variable still has significant effects. In Table 2, Models 5, 6, & 7 demonstrate that posttraining self-efficacy fully mediates the effects of training acquisition on behavioral intentions. Thus Hypothesis 2 is supported.

Table 2. Results of Path Analysis

Models	DV	IV	B	Sig.	R^2	Partial Eta Squared
1	PTSE	JI	−0.140	0.196	0.020	0.272
2	BI	PTSE	0.301	0.005	0.091	0.282
3	BI	JI	0.192	0.083	0.036	0.294
4	BI	JI	0.242	0.022	0.159	0.521
		PTSE	0.359	0.001		0.453
5	PTSE	TA	0.311	0.003	0.097	0.468
6	BI	TA	0.228	0.034	0.052	0.416
7	BI	TA	0.149	0.174	0.112	0.508
		PTSE	0.254	0.021		0.349
8	BI	JI	0.242	0.023	0.163	0.521
		PTSE	0.374	0.001		0.453
		JI*PTSE	−0.074	0.542		0.662

Note. (1) Abbreviations are provided in Table 1.
(2) All variables have been mean-centered to be N (0, 1).

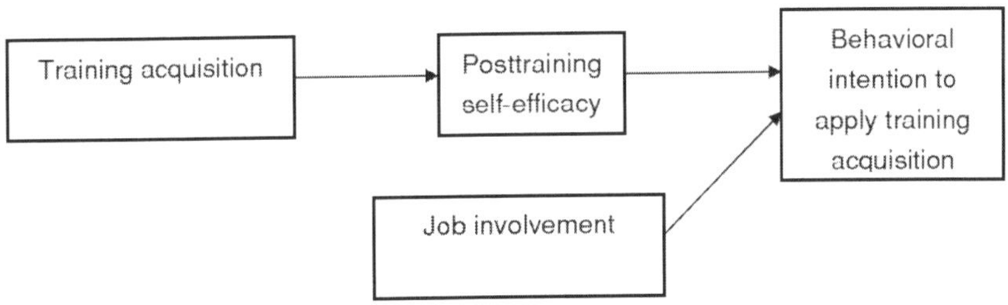

Figure 2. The tested model.

However, it should be noted in Table 2 that Model 1 indicates that job involvement does not have significant effects on posttraining self-efficacy at $\alpha = 0.05$ ($\beta = -0.14$, $p = .196$). Consequently Step 1 of the mediation testing procedure is violated. Therefore posttraining self-efficacy does not mediate the effects of job involvement on behavioral intention. However, job involvement has a significant direct impact on behavioral intention at $\alpha = 0.1$ ($\beta = 0.192$, $p = .08$). Hence Hypotheses 3a and 3b are not supported.

5. DISCUSSION AND IMPLICATIONS

In the current study, posttraining self-efficacy fully mediates the effects of training acquisition on trainees' behavioral intentions and that independent of posttraining self-efficacy job involvement significantly influences trainees' behavioral intentions.

Posttraining self-efficacy is a critical factor influencing the transfer of training acquisition into application in hospitality organizations. Posttraining self-efficacy, as noted earlier, has a learning function and a motivational function. First, the data show that training acquisition is positively associated with posttraining self-efficacy. When individuals' learning outcomes increase, trainees feel more confident in their ability to conduct the trained task. Second, posttraining self-efficacy influences trainees' application of learning outcomes. In this study, the extent to which posttraining self-efficacy affects trainees' intentions to use the training was empirically demonstrated. Posttraining self-efficacy fully mediates between training acquisition and behavior intentions, and suggests that learning will not transfer to trainees' actions until individuals feel capable and are willing to use the learning outcomes. Thus an excellent training program cannot guarantee the improvement of work; increasing both the capability and willingness to apply the learning outcomes leads to trainees' behavioral changes.

Independent of posttraining self-efficacy, job involvement significantly influences trainees' behavioral intentions to apply learning acquired in training. Previous research in training motivation suggested that the effects of job involvement on behavioral intentions would be either fully or partially mediated by posttraining self-efficacy (Colquitt et al., 2000). However, the current study does not support the mediation effects. The results from the analysis suggest that job involvement has an independent influence on trainees' behavior. When trainees have a similar level of posttraining self-efficacy, to what extent they feel their job is central and important for them is the critical determinant of transfer of training.

The idea makes sense because of the theoretical differences between job involvement and the training process. Job involvement, a dispositional work attitude, reflects the general impact of the work environment on individuals' behavior and performance and is relatively consistent over time (Steel & Rentsch, 1997). Job involvement therefore affects an individual at a more global level. The relationship between posttraining self-efficacy and trainees' behavior intentions, however, only exists within the bounds of the training process. The training process is a specific mechanism in the workplace and is not the major part (job performance) of employees' behaviors. For example, in 2000, U.S.-based organizations spent, on average, 30 hours annually on employer-provided training (Noe, 2002, p. 24). Job involvement focuses on the overall work behaviors, while posttraining self-efficacy influences training-related behaviors.

That is, job involvement and posttraining self-efficacy represent two distinct pathways influencing trainees' behavior after training. Therefore it is meaningful that job involvement affects training outcomes in hospitality organizations.

5.1 Managerial Implications

The current study suggests the critical roles of posttraining self-efficacy and job involvement in training effectiveness in hospitality organizations. It is important for hotel managers to focus on both factors within and outside the training process in order to ensure training success. After participants have been trained, hotel managers should reinforce employees' confidence in their abilities to perform relevant work roles. That is, hospitality managers should encourage employees' innate level of self-efficacy through providing opportunities to demonstrate their learning. In terms of training design, it is essential for hotel managers to understand participant-level factors in order to ensure the content of the training program is designed specifically for them. This will increase the level of self-efficacy so crucial to the transfer of training. Increasing the level of job involvement through organizational practices, such as recruitment and selection, will ensure that training costs are justified.

It is no secret that the U.S. hospitality industry depends increasingly on a workforce composed of older workers. In the case of older workers, although ambivalent, research notes that there is an age-related gradual decline in comfort with technology (Langdon, Lewis, and Clarkson, 2007). Interaction with technologies such as ATMs, front office check-in kiosks, check-in computers, and restaurant cash machines demands a greater effort from older workers than from younger workers (Jamieson and Rogers, 2000). A large part of their lack of comfort with technology is explained by older workers' inexperience with newer technologies. However, some reluctance to operate new technologies can also be explained by the fact that such systems do not provide adequate feedback (Langdon et al., 2007) and may consequently induce lower levels of perceived self-efficacy. This study has demonstrated the relationship between training acquisition (in this case, the extent to which older employees understand technology) and their perceived self-efficacy. This relationship further influences their effective use of these technologies in the workplace. The results of this study suggest it is very important for hospitality managers to pay attention to whether or not their older workers have sufficiently acquired an understanding of relevant technologies. This will be achieved through a high level of feedback on their progress through the training program. This, in turn, will influence older workers' self-efficacy and help to make them more effective. Not only for older workers, but also for younger employees, it is important that a high level of posttraining self-efficacy be achieved so that employees are more confident of themselves and their abilities. This will greatly influence turnover rates in the industry.

5.2 Limitations

The findings from this research must be applied cautiously. The current study contains a number of limitations. First, the small sample size limited the sophistication of data analyses. Although the current study has a relatively small sample, the results indicate sufficient effect sizes (partial η^2 from 0.272 to 0.662; see Table 2) to support the hypothesized effects. It is definitely desirable to collect a larger sample that will allow researchers to apply structural equation methods to test the model. Second, a student sample was used here. The students may exhibit different characteristics, such as cognitive abilities, from "real" employees or "mature" subjects. Third, the laboratory experiment may restrict the generalization of the results. The study was conducted in a simulated context of training in the use of SSTs, which may be different from the training needs of daily service encounters. Behavioral intentions measured in this study are potentially different from real workplace behaviors.

Finally, different training methods, design, and content may also lead to different results.

6. CONCLUSIONS AND FUTURE RESEARCH

The study provided adequate evidence that in the context of hospitality training, posttraining self-efficacy is a critical factor influencing the transfer of training acquisition to training application, and that job involvement has distinct effects on training transfer. Based on the conclusions, future research is needed on the following topics. First, future studies should explore the effects of posttraining self-efficacy on real behaviors in the workplace instead of behavioral intentions. Second, future research can perhaps test a "holistic" model of training motivation before, during, and after training. That would be meaningful for researchers to identify the effects across the different stages of training and for managers to design better training programs and an effective work environment. Finally, further testing on the independent model of job involvement in the context of hospitality training is needed.

REFERENCES

1. Alliger, G. M., Tannenbaum, S. I., Bennett, W., Traver, H. and Shotland, A. 1997. A meta-analysis of relations among training criteria. *Personnel Psychology*, 50: 341–358.

2. Baldwin, T. and Ford, J. K. 1988. Transfer of training: A review and directions for future research. *Personnel Psychology*, 41: 63–105.

3. Bandura, A. 1977. Self-efficacy: Toward a unifying theory of behavioral change. *Psychological Review*, 84: 191–215.

4. Baron, R. M. and Kenny, D. A. 1986. The moderator-mediator variable distinction in social psychological research: Conceptual,

strategic, and statistical considerations. *Journal of Personality and Social Psychology*, 51: 1173–1182.

5. Blau, G. J. 1985. A multiple study investigation of the dimensionality of job involvement. *Journal of Vocational Behavior*, 27: 19–36.

6. Chartrungruang, B.-O., Turner, L., King, B. and Waryszak, R. 2005. Customer satisfaction, training and TQM: A comparative study of Western and Thai hotels. *Journal of Human Resources in Hospitality & Tourism*, 5(1): 51–75.

7. Colquitt, J. A., LePine, J. A. and Noe, R. A. 2000. Toward an integrative theory of training motivation: A meta-analytic path analysis of 20 years of research. *Journal of Applied Psychology*, 85: 678–707.

8. Dabholkar, P. A. and Bagozzi, R. P. 2002. An attitudinal model of technology-based self-service: Moderating effects of consumer traits and situational factors. *Journal of the Academy of Marketing Science*, 30: 184–201.

9. Enz, C. A. and Siguaw, J. A. 2000. Best practices in human resources. *Cornell Hotel and Restaurant Administration Quarterly*, 41(1): 48–61.

10. Grau-Gumbau, R., Agut-Nieto, S., Llorens-Gumbau, S. and Martinez-Martinez, I. M. 2002. Managerial training needs: A study of Spanish tourism organizations. *Tourism and Hospitality Research*, 3(3): 234–244.

11. Haccoun, R. R. and Saks, A. M. 1998. Training in the 21st century: Some lessons from the last one. *Canadian Psychology*, 39: 33–51.

12. Hair, J. F. Jr., Anderson, R. E., Tatham, R. L. and Black, W. C. 1998. *Multivariate data analysis (5th ed.)*, Upper Saddle River, NJ: Prentice Hall.

13. Jamieson, B. A. and Rogers, W. A. 2000. Age-related effects of blocked and random practice schedules on learning a new technology. *Journal of Gerontology*, 55B(6): 343–353.

14. Jones, G. R. 1986. Socialization tactics, self-efficacy, and newcomers' adjustments to organizations. *Academy of Management Journal*, 29: 262–279.

15. Kim, B.-Y. 2006. Managing workforce diversity: Developing a learning organization. *Journal of Human Resources in Hospitality & Tourism*, 5(2): 69–90.

16. Kirkpatrick, D. L. 1996. "Evaluation". In *The ASTD training and development handbook ed.*, 4th, Edited by: Craig, R. L. 294–312. New York: McGraw-Hill.

17. Kraiger, K., Ford, J. K. and Salas, E. 1993. Application of cognitive, skill-based, and affective theories of learning outcomes to new methods of training evaluation. *Journal of Applied Psychology*, 78: 311–328.

18. Kramer, S. H. and Rosenthal, R. 1999. "Effect sizes and significance in small-sample research". In *Statistical strategies for small sample research*, Edited by: Hoyle, R. H. Thousand Oaks, CA: Sage.

19. Langdon, P., Lewis, T. and Clarkson, J. 2007. The effects of prior experience on the use of consumer products. *Universal Access in the Information Society*, 6: 179–191.

20. Lee, Y.-K., Nam, F.-H., Park, D.-H. and Lee, K. A. 2006. What factors influence customer-oriented prosocial behavior of customer-contact employees?. *Journal of Service Marketing*, 20: 251–264.

21. Noe, R. A. 2002. *Employee training & development (2nd ed.)*, Boston: McGraw-Hill/Irwin.

22. Pollitt, D. 2006. Red carnation blooms through management training. *Human Resource Management International Digest*, 14(6): 11–14.

23. Reeve, C. L. and Smith, C. S. 2000. Refining Lodahl and Kejner's job involvement scale with a convergent evidence approach: Applying multiple methods to multiple samples. *Organizational Research Methods*, 4: 91–111.

24. Salas, E. and Cannon-Bowers, J. A. 2001. The science of training: A decade of progress. *Annual Review of Psychology*, 52: 471–499.

25. Steel, R. P. and Rentsch, J. R. 1997. The dispositional model of job attitudes revisited: Findings of a 10-year study. *Journal of Applied Psychology*, 82: 873–879.

26. Tracey, J. B., Hinkin, T. R., Tannenbaum, S. and Mathieu, J. E. 2001. The influence of individual characteristics and the work environment on varying levels of training outcomes. *Human Resource Development Quarterly*, 12: 5–23.

27. Tracey, J. B. and Tews, M. J. 1995. Training effectiveness: Accounting for individual characteristics and the work environment. *Cornell Hotel and Restaurant Administration Quarterly*, 36(6): 36–42.

28. Von Bergen, C. W. and Mawer, W. 2007. Recouping training and development costs using preemployment agreements.*Employee Responsibilities and Rights Journal*, 19: 127–143.

29. Zhao, X., Zhan, J. and Namasivayam, K. 2004. Factors affecting training success in China. *Journal of Human Resources in Hospitality & Tourism*, 3(1): 89–105.

3

ROOM SERVICE PRINCIPLES AND PRACTICES: AN EXPLORATORY STUDY

Stanley Douglas Suboleski

University of Nevada, Las Vegas

ABSTRACT

Room service is the delivery of food and beverage products to a guest's sleeping room in a hotel. It is an important, distinguishing characteristic and a necessary service for firstclass and luxury hotels. Very little academic research has been conducted on the operating procedures of room service. This was an exploratory study that aimed at identifying the key principles and practices of room service operations in first-class and luxury hotels. This study used a mixed-method approach. Using content analysis on operating audits for room service from a range of hospitality companies, expert panel review, field testing, and analysis of variance (MANOVA), this study resulted in a generic room service audit representing the key principles and practices currently in place in the first-class and luxury hotel segment. Key elements of the generic audit were then tested in a sample drawn from a population of hospitality professionals throughout the United States, resulting in what they determined to be the most important elements in room service from the abbreviated list of items that was presented to them.

KEYWORDS

room service, operations, first-class hotels, luxury hotels, hospitality, operating audit, policies and procedures

PART I

1. INTRODUCTION

Room service is an important component of the overall service experience within upscale hotels and resorts. Also referred to as "in-room dining," room service is defined as an operating department within a lodging facility that takes guest orders for food, prepares that food, and delivers it to the guest's room for consumption there. Although some room service operations function in a capacity that supports banquets or convention services departments with food delivery and services to guests throughout the hotel, this study intends to specifically focus on room service to the guestroom itself.

Two widely utilized hotel rating systems, AAA and Mobil, consider in-room guest dining to be one of the more important aspects of hotel services. For example, no hotel can get the top diamond or star ratings/awards from these organizations if its operation does not have a 24-hour room service operation in place (Ninemeier & Purdue, 2008). In hotel operations, room service has long served as a convenience to the guest. Room service provides an opportunity for guests to eat at their own pace, in their own room, whether it is for practical, time-saving purposes or for luxurious indulgence.

Menu prices for room service are often at a premium in comparison to the hotel's restaurant menu offerings, which are frequently the source for the room service menu itself. Delivery of food and beverage to a guest room is generally more labor intensive than restaurant services, if for no other reason than less distance needs to be covered to successfully deliver meals in a restaurant, as opposed to taking that meal to a hotel guest room which may be quite distant from the kitchen where it was prepared. Hotel guests who utilize room service are usually aware of the additional expense of ordering food delivery and generally accept the premium cost of convenience as an aspect of their bill. If developed and executed in a fashion that is appropriate for guest needs, hotels may generate significant revenue at potentially healthy profit margins from their room service operations.

It can be argued that in-room dining is not only an important revenue center, but also that it has become more so due to the decline over time of other traditional in-room guest revenue generators. In recent years, advances in technologies appear to have provided guests with the ability to avoid charges that were long common to the lodging industry. The evolution of personal telecommunication technologies has adversely impacted the telephone revenues a hotel generated in the past. A great portion of the population has cellular phones that allow them to bypass in-room access fees and marked-up time charges for calls. The same has begun to happen with in-room entertainment. The proliferation of portable multimedia devices may allow guests to bring their own content with ease to a guest room, avoiding traditional alternatives, such as in-room movie fees.

As for additional food revenues, many full-service hotels cannot or do not control the flow of room delivery of food and beverage from outside sources. It is difficult to control these outside deliveries logistically, not to mention that restricting a guest's free use of choices of service may impact the overall perceptions of the quality of that guest's stay. With other sources of revenue falling, the hotel industry needs to look at the overall room service process, from both a functional fashion, focusing on the efficiency and quality of the operation, as well as from a marketing perspective, focusing on guest needs, expectations, and perceptions.

However, in many operations, room service is often perceived to be disorganized, late, unprofitable, inconsistent, poorly promoted, and under-utilized by guests. This is a problem. Thus, if a hotel sees room service as an important aspect of its operation, management needs to work at promoting its usage and enhancing its quality (Ninemeier & Purdue, 2008).

Currently, there appears to be no agreement on the best principles and practices to utilize in the development and operation of an effective and profitable room service department. There are no concise academic articles or textbooks that discuss how to best conduct room service operations. Yet, all major lodging organizations have room service departments that, like any other department, have standard operating policies, rules, regulations, and service standards. Firms in the lodging industry do not currently share their ways of operating room service with each other, perhaps due to the desire to protect what may be seen as a point of competitive distinction. Without a comprehensive overview of the various ways in which hotel companies design and execute their own inroom dining efforts, it is difficult to say which are most effective.

Research needs to be conducted to identify common operating trends that lodging companies share with regards to room service, as well as to determine unique operating processes that can be more widely adopted in the industry. There are great opportunities for a lodging organization to increase its return in this department if the property or parent company can understand the drivers for consumption, identify guest attitudes towards room service, and most importantly, design an effective set of operating procedures and controls to provide consistency, value, and quality in the guest in-room dining experience. This study examined room service operations, service standards, and techniques; as such, it represents the first investigation of its kind to examine standards for service pertaining to hotel room service.

1.1 Purpose of the Study

The purpose of this study was to identify the standards for in-room guest dining that currently exist in first-class and luxury hotels. For this research, a first-class and luxury hotel was defined as an upscale hotel which typically costs more than the average accommodation, and which offered a combination of style and a range of personalized services not usually found in average lodging properties (Ninemeier & Purdue, 2008). This was an exploratory and mixed methods study that sought to identify the majority of the varying principles and practices used by the lodging industry for in-room dining. This study focused on answering several related research questions, including the following:

- What are the most vital and salient characteristics of a successful, full-service, inroom dining program?
- Which operating processes can be identified as foundational practices necessary to ensure a quality experience for the guest?
- Is it possible to create an overview of these identified practices that would address all, if not most, types of first-class and luxury hotels?
- What are the noteworthy, structural differences between these standards across company or asset types?

The goal of this research was to develop an overview of these best practices and to ultimately generate a generic operational audit of these commonly used procedures. Reflecting on the practices of this segment of the industry will allow further research into which components are most effective, distinct, or strategic. This, in turn, could assist a single property or a hotel chain in maximizing the potential of this important revenue center.

The academic significance of this project is to further the understanding of how a hotel designs and executes its room service offerings. This work replicated and extended the works of Moreo and Savage (1990), Moreo, Sammons, and Savage (1997), and Wood, Moreo, and Sammons (2005), who examined hospitality operational audits in accounting, front office operations, and housekeeping and food services, respectively, and added another functional department to those previously analyzed. In addition, this research enhanced the processes of operational analysis that preceded it by adding a mixed method component to the creation of a generic audit, thus allowing the resultant audit to be reviewed and verified by industry professionals that execute such tasks in room service regularly. The development of a comprehensive operating audit for room service that focused on both process and outcome augmented the generation and application of quality control techniques.

The practical significance of this research is to assist hotels in analyzing what is being done in the industry with regard to room service, from how it is conducted, to its preparation and delivery. This perspective allows lodging properties to examine this function in a more globally understood context, while still focusing on quality and distinction. Beyond guest perception of the delivered product (outcome), an operational audit allows the operation to test itself objectively by measuring the process as it is completed against the standards it was supposed to achieve. A better understanding of room service assists a hotel in selling more in this profit center and generating better food and beverage revenues.

An overview of the literature was conducted with the intention of identifying key features of the process that could be matched with the data collected in the field later in the research process (Locke, 2003). Many of these features were developed into codes for the coding process of the audits. In most cases, key-operating practices were noted and discussed contextually within the literature review and then identified in as part of the coding development process in this study.

The literature review examined a variety of works that related information about the operating trends, procedures, and structures of a RS department. Literature was scarce on specific policies and tended to be very general, but assisted in the development of hypotheses for this study. The fourth research question was concerned with differences in RS operations based on structural differences in the lodging operation. Kandampully (2007) noted that the larger an operation, the more a service organization should attempt to standardize services in order to reduce variability. Vallen and Vallen (2008) established some of the traditional categorizations of lodging operations. Among them was the size of a hotel, which was based on the number of rooms in the facility. This identified size as an independent variable that might affect the way in which a RS department was operated. Another common categorization of hotels was the type of facility (Vallen & Vallen, 2008), which was determined by the kind of services offered, and the customers that were targeted. If services available varied among property type, then the category of hotel was another independent variable that should be investigated. In Buick's study (1988), only a small portion of the properties had a RS department that was free standing. The majority of hotels in the study had a RS function, but operated it out of an existing restaurant. The availability of employees to perform RS duties could be affected by the workload they had in the outlet in which RS was being conducted. The absence of presence of a free-standing RS department was another independent variable identified for this study that might have an impact on the manner in which the services were offered and produced. Kandampully (2007) indicated that franchised operations had less freedom in making operating decisions due to policies mandated by the corporate offices of the franchise brand that were aimed at standardizing geographically distant facilities. The creation of such a brand image is essential to a franchised operation, but it is unclear whether such policies have an impact on the way in which RS is operated. The franchise status of a facility was identified as another independent variable to investigate in this study. These categorizations of hotel types assisted in the development of four hypotheses for this study:

H1: There is no difference in the perceived importance of key room service practices between employees of a large hotel and employees of a small hotel.

H2: There is no difference in the perceived importance of key room service practices between employees of a resort hotel and employees of a non-resort hotel.

H3: There is no difference in the perceived importance of key room service practices between employees of a hotel with a free-standing room service department and employees of a hotel without a free-standing room service department.

H4: There is no difference in the perceived importance of key room service practices between employees of a franchise hotel and the employees of an independent hotel.

PART II

2. LITERATURE REVIEW

The literature review focused on three distinct areas: lodging room service operations, service quality and quality control, and both operational and hospitality audits.

This work provided an overview of room service literature in both academic and trade journals, as well as in textbooks. Specific manuscripts that focused on the execution and operation of room service were sparse. Although there was no comprehensive piece on the topic, there were manuscripts that helped to identify practices that were developed and seen as effective. There were also pieces that describe specific issues or problems with room service operations and solutions that were applied to remedy them.

The study also examined the literature on service quality and quality control processes. This included an overview of quality control in hospitality, its importance, history, and applications. Focus was on the necessity of such programs, and the benefits they produced in process and outcome of a service operation.

In addition, this review examined the literature on operational audits, in both theory and application. This included the history, benefits, and protocol of audits. Focus was on both how and why an operational audit was developed, as well as on how it was utilized. The effectiveness and vitality of the operational audit process was an important point, as it drives and supports the purpose of this study.

The overall literature review synthesized the three key areas of room service procedures and issues, operating audits, and service quality, and provided support for the importance and relevance of this specific study.

2.1 Room Service

This section of the review was the most specific and was concerned with literature regarding in-room guest dining. The body of literature regarding room service was sparse, but that which was located was quite helpful. None of the work was specifically foundational in terms of theory, as most of the literature was from textbooks, trade or popular press, and training manuals. The goal was to discover what the literature said about room service and current trends, critical issues, and practices.

2.2 Description and definition.

Cichy and Wise (1990) described room service (RS) from a guest perspective. They defined RS as a guest's ability to order and enjoy food and beverage in the privacy of his or her guest room. Ninemeier and Purdue (2008) created a very simple and straightforward definition for room service from an operator's perspective as the act of serving food and beverage products to guests in their sleeping rooms. For the purposes of this study, the working definition of RS was considered as the operational process of taking the order, preparation, and delivery of food and beverage products to a guest sleeping room as an act of enhanced service in a hotel.

According to Rutherford (1990), as the first-class and luxury hotel industry became increasingly competitive in the late 1980s, providing full-service, in-room food and beverage delivery 24-hours a day became a necessary competitive amenity, and most hotels had to scramble to improve guest room delivery services. However, at many hotels, RS operations were neglected, and subsequently, RS was neither planned nor designed well in new hotel construction. Hanson (1986) added that RS was a hotel amenity expected by seasoned travelers, but noted that, considering the setup, distribution, and collection of room service trays and carts, as well as the cost of maintaining a staff to respond promptly to guest orders, RS was by definition an inefficient operation that adversely affected productivity. Turk (1979) elaborated by stating that RS was seen as a necessary evil. It was seen as necessary because some guests prefer to eat in their rooms, and evil because of the high labor costs and low volume. He also noted that RS was a distinctive and unique competitive feature that the hotel could offer to the guest and should be treated with an according degree of importance.

As a guest service, RS tended to be quite lightly utilized. Wolf (1992) discovered that approximately 4% of business and leisure travelers ate RS breakfast. There was literally no market for RS lunch, and only about 2% of business travelers ate RS dinner, while leisure travelers did not patronize RS for dinner (Wolf, 1992). RS may be utilized for specific purposes. RS guests were often seeking either status, convenience, or a special touch or feeling to a special occasion, RS operations should look at its clientele and determine whether the appropriate positioning of the amenity is for its functional or its symbolic merits (Cichy & Wise, 1990). Functional merits included privacy, security, solitude, and convenience, while symbolic merits encompassed novelty, prestige, and special occasions. Some guests purposely selected hotels based on the availability of RS, while other guests selected properties on the basis of travel and lodging rating systems, such as AAA and Mobil. The highest ratings by these organizations required the presence of 24-hour RS operations (Ninemeier & Purdue, 2008). Ratings are generated by announced and unannounced personal visits to the property by professional inspectors who evaluate the hotel's services utilizing standardized checklists. Additionally, in order to be represented by certain lodging affiliations such as Small Luxury Hotels or Preferred Hotels and Resorts, first-class and luxury properties must offer RS.

Turk (1979) viewed RS as the great-untapped potential of food and beverage operations, and noted that, while demand for 24-hour service was not overwhelming, it could be profitable. Wolf (1992) identified reasons why guests shy away from RS. The two largest reasons were high prices and slow service, accounting for 45% of all guest complaints regarding RS. Other salient reasons for guest adversity included food quality and a limited or unappealing menu. A study by Mohsin (2007) focused on an importance performance assessment of RS operations. His study supported Wolf's and identified the largest gaps in RS as the quality of food, the prompt response during ordering, and the value-for-money. In fact, not a single factor in RS matched the guest perception of performance with their perception of importance, indicating that there were great opportunities in the development of a functional RS operation that was geared towards guest satisfaction.

Cichy and Wise (1990) identified the types of hotel guests that did utilize RS and the primary reasons for using it. Those users included business travelers, who were not necessarily price-conscious because they were not using their own money to pay for the services. It also included bus-tour groups, who were interested in timing and convenience, and sports teams, who tended to eat individually on their own schedule and timing. Other guests using RS included international guests, who may not be comfortable with language and interaction with employees from another culture, families with children, who were seeking a quiet, controllable atmosphere, and guests with disabilities, simply seeking convenience.

2.3 Overview

A review of operational hospitality textbooks specifically in hotel and foodservice management provided detailed overviews of the room service process on an idealized basis, without specificities or differentiations related to brand that would likely be present in the operating audits used for data analysis. Davis and Stone (1985), Rutherford (1990), Cichy and Wise (1990), and Ninemeier and Purdue (2008), have all authored hospitality operations textbooks that review the room service process in detail. In addition to providing a thorough overview and description of the RS operation, information gathered from these texts assisted with the initial coding process conducted later in the study.

Cichy and Wise (1990) identified the guest components of successful room service. These factors should be present when designing a RS operation or process that is focused on guest satisfaction. They noted that, while RS should be focused on guest needs, it should also have its eye on the materials and resources of the hotel, and their potential impact on quality food delivery to guest rooms. These components included prompt and courteous responses from staff when placing orders, correct orders, fast delivery, tact and courtesy from RS food-deliverers, safety practiced with equipment used in the guest room, hot and cold foods served at appropriate temperatures, and finally, prompt tray or cart removal upon completion. Identifying important guest needs and wants and including them in the operational design of the department was a recommended strategic starting point for the design of a typical RS operation.

2.4 Process components.

To provide a digestible overview of the idealized process of the RS function, the literature from operational textbooks (Cichy & Wise, 1990; Davis & Stone, 1985, Ninemeier & Purdue, 2008; and Rutherford, 1990) was reviewed and incorporated into this dissertation. While no one process was accurate in describing the myriad ways that a hotel could choose to manage a RS operation, the functional steps necessary in executing RS business can be identified in very large areas or categories. It is these areas that became the foundational codes for the generic audit development. In the methodology section, a more detailed discussion will be held regarding which slight variations in the general processes demanded specific coding and categorization, and which could be grouped together due to inherent similarities.

In general, based on the references cited in the previous paragraphs, there were six "sections" to the process of handling a RS order from a guest room. These included pre-preparation, order taking, order routing, order preparation, order delivery, and postorder activities. Following is a description of the components contained within each section.

2.5 Pre-preparation

The first section in the operating processes of RS was called pre-preparation. Pre-preparation is involved in ensuring that adequate information, supplies, materials, and resources are available and on-hand for use in the RS department. Davis and Stone (1985) identified that the process of preparing for a shift in RS needed to begin with adequate, accurate information. They identified important data in assisting the RS department in planning, staffing, and reserving equipment and materials needed for a shift. Data, often supplied from the front office of the hotel that assist the RS in pre-preparation included: occupancy levels, group business currently in-house, the marketsegment mix of current and arriving clientele, banquet business and timing, and VIP or amenity distribution. From that information, RS needs to be ready to act. It must ensure that it has adequate staffing levels to meet the potential demand.

There were also requirements for specific operating inventory (Rutherford, 1990). Inventory requirements included trays, various types of carts, candles, tablecloths, napkins, utensils, decorations, and other relevant specialty items for the guest table/cart. This operating equipment should be stored close to the work area for easy access. Whenever possible, based on available information circulated in the hotel, RS should use preset trays and carts prepared in advance the shift prior (Cichy & Wise, 1990). Additionally, as a general rule of thumb, RS should identify, establish, and maintain set par-stock levels on all items needed to accompany food delivery, and refresh to that par level daily or by shift (Davis & Stone, 1985). All preparation of this kind done in advance benefits the operation by allowing it more time to react to other un-planned or unexpected variances in business levels.

2.6 Order taking

The next step in the process was the order taking. This is where the guest decided to purchase food and beverage products for guest room delivery and initiates the process by contacting RS, usually over the in-room telephone. Cichy and Wise (1990) described the process as a sequential series of events, each dependent on the step before it for maximum quality and efficiency. Politeness and appropriate service attitude and etiquette were vital during this step of the process, as it set the tone of the guest encounter, influencing the guest's expectations of the service experience, either negatively or positively.

A specific script should be developed that is appropriate for the guest segment(s) being serviced. The telephone should be answered within a certain number of rings and with a certain scripted greeting and tone of voice. The attendant taking the order should identify himself or herself by name, and that the guest has reached the RS department. The attendant should also collect the guest name and information from either an electronic caller-ID device and/or by verifying with the guest directly. Attendants should be knowledgeable and well trained enough to answer specific questions about dishes as they arise. Attendants should be trained to recognize and utilize opportunities to up-sell or cross-sell products and services from RS and the rest of the hotel during the conversation. The attendant should take the order by whatever system the hotel uses, be it manual or computer-driven. The attendant is responsible for garnering all the relevant details about the dishes, inclusions, omissions, or special requests of the guests, and should always maintain a back-up paper trail of the order. After repeating the order to the guest, the attendant should estimate the delivery time and thank guests for their patronage.

In addition to this sequential overview, Rutherford (1990) noted that the RS department should always use a standardized RS ordering form. Cichy and Wise (1990) stressed that telephone etiquette is of paramount importance and all attendants should be trained on the proper way to ask a guest to be put on hold when call volume to the RS department is high. Davis and Stone (1985) also believed that guests deserve a return call to advise them if the original timing estimate for delivery looks like it will not be met.

2.7 Order routing

After the order has been taken and reviewed with the guest, it needs to be routed to the preparation area in order to start production of the food. Routing the order involves the order-taker delivering the order on an appropriate form to a production area, or giving the order to RS delivery personnel to do so. Cichy and Wise (1990) noted that the specific process is dependent upon the technology available at the facility. If there is a Point-of-Sale system (POS), then the order-taker or delivery person can take the guest order and enter it into the system, where it will be communicated directly to the production staff.

In manual systems in facilities where no automated system is present, a person may need to physically carry the order the production area. After entering the order into whatever system exists, the order should also be entered into a RS order log so that the status of preparation and delivery may be checked during operations, or reviewed as an analysis tool by management at a future date. An efficient RS operation should also have an ordering system that allows for guest orders to be sent to multiple production facilities, such as a bar/lounge for drinks and a restaurant for food (Rutherford, 1990). Finally, a copy of the order needs to be delivered or processed with a cashier or an electronic system to ensure payment is procured.

2.8 Order preparation

Once an order was taken and routed, the production phase could begin. In the production phase, there were components of this process that may be the responsibility of the RS deliverer, as opposed to the kitchen production staff. For this reason, it may be necessary to consider the food preparation as a separate function from the order preparation in the coding process. There was no literature on the preparation of food from a culinary perspective in a RS operation. However, certain protocol may be established in a facility to delineate responsibilities for cooking the food, as opposed to preparing the cooked food for plating, presentation and delivery. The operating audits utilized for this study guided the level of separation and detail between food preparation and order preparation and addressed both the RS server's responsibilities in "finishing" food items as well as the cooks' responsibilities in preparing the items.

It is imperative that any such kitchen policies and procedures were clearly established for the preparation phase so that there were no missing ingredients for the guest order (Davis & Stone, 1985). RS attendants needed to be kept aware of which dish components were the responsibilities of the delivery staff, and which were the responsibilities of the production staff. The RS attendant should prepare any support or side dishes promptly as required. During the actual food preparation, the RS attendant should prepare a fresh tray/cart with materials needed to consume the order, or utilize a preset tray/cart. In either case, the tray/cart should be reviewed in detail in order to ensure all needed materials are present. The RS attendant wraps, caps, or lids hot and cold dishes to hold their temperature as they are received from the production facility. This may involve multiple RS employees going to various production outlets to assemble the components of the guest order, so it is important to keep awareness and timing in mind. Many hotels had a firm policy similar to that in a dining room where an expediter examined the tray and the order as a review prior to releasing it for delivery. This order checking could help to reduce return trips when an item is forgotten (Ninemeier & Purdue, 2008).

2.9 Order delivery

Once the order has been reviewed, approved, and released, it is time for delivery to the guest room. Cichy & Wise (1990) established detailed service guidelines for this phase, considering it to be quite important in terms of effect upon guest perception of quality and satisfaction. Food should be delivered as quickly as possible, using the fastest route to the guest room. Food temperature is an important element to the guests, so food that is ready should be taken immediately or held appropriately until the entire order can be delivered.

The RS staff should use service elevators, hallways, or dumbwaiters wherever possible to avoid guest traffic (Rutherford, 1990). All delivery personnel should be trained on proper cart operation and tray handling to avoid delays caused by accidents or spilled product. RS delivery personnel should be trained to use uniform greeting and identification at the door of the guest room. Upon entry to the room, the delivery person should ask if any assistance is desired in the setting out or displaying of the order. If desired, a delivery person should be well versed and adroit at tableside presentation of items in a style appropriate to the hotel and its guests. Delivery personnel should be conditioned to be very aware and discrete when it comes to guest privacy. No personal items of a guest should be touched or moved without first asking permission (Rutherford, 1990).

Delivery personnel should ensure cart wheels are locked and heating elements or open flames in carts are extinguished. Foods should be thoroughly described, as well as proper use of any equipment left behind in the guest room. Additional assistance should be offered as a closing act, along with information provided to the guest as to the process in place to get a cart or tray retrieved from the guest room or hallway. After departure, on the way back to the RS area, delivery personnel should scan the hallways for a tray or cart to take back with them, in order to reduce clean up time, keep the hotel looking fresh, and avoid running out of needed trays or carts in a case of an unexpected rush. Davis and Stone (1985) also suggested having a system in place to check on the order with the guest after it has been delivered, thus providing a chance to rectify any missed items or preparation errors, and to remind the guest of tray or cart return processes in place.

2.10 Post-order activities

Post-order activities included things that were done after the delivery has been made. Cichy & Wise (1990) indicated that a strong income control procedure needed to be implemented. This system would help to ensure that all charges are correctly entered into the system and posted and that the proper method of guest billing and collection is employed. Special care should be taken with the handling of cash in order to reduce employee theft. The RS department should have either an electronic system or a communications policy with accounting or the front desk (if there is no direct system), which allows verification of a guest's ability to charge a RS purchase to the room. Wolf (1992) added that it is wise to have a system in place that solicits and records all feedback from RS guests. It can be quite important to follow up with a phone call to the guest room a few minutes after order delivery in order to ensure that no item was missing from the order or that the guest does not require any additional service to complete the consumption of the meal. That strategy should not stop with the collection of the feedback. Rather, it should continue on into planning strategically to design systems that prevent service failure in the future, given similar circumstances.

2.11 Challenges

There were many challenges that made it difficult for hotels to properly and efficiently deliver foodstuffs to guest rooms. In many cases these challenges were universal to the industry and to RS itself. While it was vital to identify these challenges, not all were operational in nature. The focus of this study was on operational practices that impacted efficiency and quality in a RS operation and to subsequently develop an audit that will help readily identify the adherence to these practices.

The chief challenge in RS seems to be that of profitability. Ninemeier and Purdue (2008) discussed that RS losses were due to high labor costs. It took a lot of time to transport food to sometimes-remote areas in the hotel. Additionally, inventory costs for warming carts and delivery trays were high. Costs related to glassware, flatware, and utensils were high because products were out of the hotel's direct control for long periods of time and breakage, loss, and theft often occurred (Rutherford, 1990).

Cichy and Wise (1990) indicated that very few RS operations were profitable. RS will often cost even more if it has its own separate production facility, as opposed to operating within an existing, 24-hour restaurant's kitchen. Hanson (1984) noted that RS is expensive to operate. Turk (1979) added that RS departments were often notoriously undermanaged and patrons underserved with a general lack of urgency or poor service culture. He noted that RS generally failed at providing at least the same level of quality service as is existent in the hotel's restaurants. Ninemeier and Purdue (2008) stressed that, even with high prices, many RS operations were not profitable and that hotels tended to view RS as a guest amenity as opposed to a revenue center.

Breakfast was a big meal for RS. According to Cichy & Wise (1990), breakfast was the most popular meal, with upscale hotel RS business accounting for 35% of all breakfast business. However, breakfast was a difficult meal due to time demands and menu items (such as eggs or toast) not holding their heat well during transport. Viable product substitutes should be considered and used, where appropriate.

Cichy and Wise (1990) identified issues with guest room design that may adversely affect RS consumption or perception. They noted that any design of a guest room must include space to roll in a cart. If there is a table cart, then chairs should be at appropriate height to use at the table cart. At the very least, the in-room table should be large enough to accommodate two diners and their dishes. In many cases, hotels did not meet these simple requirements, ones that would make dining more pleasant and less of an inconvenience to a guest deciding to eat a meal in a guest room.

Finally, Ninemeier and Purdue (2008) identified several other challenges facing a full-service RS department. These included communication issues, such as language barriers with international guests and incomplete or unclear orders. By the nature of the service interaction between staff and guest, there were lost opportunities for up-selling product in RS compared to the way that it could be done in a restaurant. In the guest room, the employee generally only gets one chance to remedy a service deficiency with ease, due to the distance between the production and the consumption portions of the RS experience. This makes it vital to have well-trained, knowledgeable staff that can set up the meal properly and as needed, open wine bottles properly, and give distinctive and friendly service with a pleasant attitude.

2.12 Opportunities

Though there were many challenges and difficulties facing RS, it would be unfair to characterize it solely as a necessary but unwanted feature saddled upon the operating staff of the hotel. Finn (1977) noted that hotels must embrace RS as a guest convenience that can add revenues and profits to the operation, asserting that breakfast, as the busiest meal period in a hotel was a huge revenue opportunity. Cichy and Wise (1990) agreed, and added that RS breakfast was the number one guest service for which customers were willing to pay extra. They added that breakfast was not the only opportunity that RS could enjoy. Other areas where revenue opportunities were available included but were not limited to, bottled wine and specialty liquor service, guest amenity trays and baskets, executive/VIP coffee and break services, host bar set-ups, and hospitality suite parties and meetings. Davis and Stone (1985) added that some operations with hospitality suites will transfer the food and beverage responsibility for hosted meetings in a guest room from the banquet department to the RS department in order to offset operating losses experienced by necessary service staffing minimums. Turk (1979) pointed out that hotels could compensate for high labor costs by maintaining efforts to increasing RS volume through marketing and word-of-mouth. Hubsch (1966) stated that one of the best opportunities RS could provide was that a successful operation could become a drawing card for guest room sales, the primary business focus for the facility.

2.13 Organizational structure

Often, the size and organizational structure of the RS department differs from property to property (Cichy &Wise, 1990), with some facilities utilizing existing restaurant staff to take orders and deliver them and some having separate freestanding departments. Regardless of the demand or reasoning, it is vital that the organizational structure and inherent responsibilities of positions in that structure are clearly established in a hotel. A typical organizational structure for a RS department would include a RS manager and possibly an assistant RS manager at the uppermost levels, a RS captain in a middle supervisory position, and RS order-takers, delivery attendants and bus persons as the front line of guest contact and service provision (Cichy & Wise, 1990).

Ninemeier and Purdue (2008) noted that some larger facilities separate the RS function from the restaurant and main kitchens entirely, employing a kitchen manager/cook to supervise back-of-the-house food preparation responsibilities. They also noted that the RS mangers are an important position in the structure, not only because they supervised the daily operation, but additionally, they were charged with the important task of RS menu planning and design.

Cichy and Wise (1990) developed a list of important competencies for RS staff to possess. They believed that RS font line staff should be specifically trained to have high levels of menu product knowledge, including pronunciation, specials and signature dishes, ingredients, and preparation methods. They should also be empowered to a degree to offer alternatives or attempt to accommodate special dietary needs or restrictions without unduly lengthening the process.

2.14 Academic Research

There are few academic studies on room service. A key article by Buick (1998) discussed the important operational aspects of providing room service. It reviewed how room service was organized in 29 Edinburgh, Scotland, hotels. It examined which of the upscale properties had room service 24-hours a day, and whether the room service department was organized as a separate department. Of the 29 operations examined, 96.4% had 24-hour room service available (Buick, 1998). The study examined where in the facility the department was located. Results showed that only

22.2 percent of the room service operations had their own department, while the remainder was assigned to other food and beverage operating departments (Buick, 1998). The study also revealed that the major point of contact for a room service exchange was the guest room telephone. It also disclosed that the majority of meals served, around 90 percent, were served during the breakfast period. Although the study offered little about the actual execution of service preparation and delivery, it provided a background of room service in an operational context that will help paint a better picture to readers of the complexity of the operation.

A study by Mohsin (2007) reviewed customer perception of satisfaction in Australian hotels. The study focused on the front office, room service, and restaurant operations. Though this study did not focus on customer perceptions, the survey instrument indicated a few operational areas that were of importance to the guest with regards to room service. Of specific importance to guest perception in room service were prompt responses from order takers and prompt delivery of product. Again, this literature review section's goal was to ascertain what the industry was doing as well as should be doing to ensure quality outcomes, and such supportive information ensures that these two constructs will be considered within the coding scheme of the content analysis, and revisited as needed in a panel or focus group.

One academic article by Eaton (2004) discussed the risks of sexual harassment to hotel employees who delivered food to guest rooms. The piece took a decidedly legal perspective in terms of its tone, but generated several suggestions on how a hotel organization should structure food delivery to protect its staff and itself from the legal, emotional, and financial ramifications of sexual harassment. While there does not appear to be extensive discussion of such risks or practices, its limited presence in the literature was an indicator of its potential inclusion in a checklist of operating practices.

2.15 Marketing

Part of a successful RS operation is its internal marketing and cross-selling efforts within the property. Cichy and Wise (1990) stressed that it was essential for the hotel to drive traffic to the RS department by using internal promotional and advertising strategies. For example, the sales department should have a fact book with relevant information on products and services in the hotel. In that collateral, RS needs to be positioned as a benefit and attractor to potential groups. In daily operations, frontline employees should use any opportunities during guest interaction, such as wake-up calls and greetings, to sell RS.

Hubsch (1966) indicated that well merchandised food and beverage operations were a necessary part of a hotel image, especially in the "hidden" RS department. RS sales information should be prominently displayed via advertising and in-room collateral, including but not limited to posters, flyers, table tents, brochures, and in-room TV ads. Cichy and Wise (1990) added that elevators are prime high traffic areas for posting of RS information.

Creative marketing should be part of a RS manager's duties (Turk, 1979). Merchandising and positioning are important for RS, and if treated properly, the hotel can sell RS to the guest as a kind of bonus the guest can elect to treat themselves to when traveling. Better merchandising might include more elegant table settings, smartly dressed and well-trained, uniformed staff, and graphically interesting menus. Not that all marketing in RS needed to play to the upscale experience, it could also play to price as an attractor. Cichy and Wise (1990) also suggested that RS managers should consider discounting orders to drive business or consider having a program where multiple purchases receive a lower price.

2.16 Menu Engineering

The RS menu design should be considered as one of the more important aspects of the RS sales and operations processes. They are the chief point of information for guests that might spur them to make a purchase. Of course, RS menus should reflect the operation's products and services in alignment with customer expectations and needs (Cichy & Wise, 1990). Menu planning was considered important. Not only did menu items need to meet quality and preference criteria for the guests, but also needed to meet profitability requirements for the operation.

Generally, RS menu engineering produces higher prices and lower varieties than their counterpart restaurant menus. Some operations chose to offer special and unique dishes not on menus in hotel restaurants, even including fast-service items to compete with outside entities available to the customer (Cichy & Wise, 1990). Additionally, RS menus were intentionally designed to contain items that can maintain their quality during transportation to a guest room. Some hotels only offered items that they could prepare and deliver with consistent quality on their RS menu, limiting the size to just a few items per meal period. Some offered only the most popular items from their other food and beverage outlets. Some hotels have opened RS kitchens that only prepare a single type of item such as pizzas or sandwiches (Cichy & Wise, 1990).

Ultimately, Cichy and Wise (1990) stressed that RS managers should use in-depth menu engineering practices to select and monitor the success of menu items. This process included analyzing the demand and the profit contribution of each dish on the menu, and adjusting prices or adding and removing items as warranted by the analysis. They developed a list of salient objectives to consider when designing a RS menu, which included cook times, dish popularity, product portability and temperature, profitability, menu size and related resources needed, and hotel brand or segment requirements.

In any design effort, the menu should reflect the property's image. Also, it should be easy to read, easy to understand, and easy to locate dishes (Ninemeier & Purdue, 2008). The RS phone number should be readily identifiable and prominently displayed. The RS menu location should be easy to find in the guest room. Multiple languages and menu item photos should be available for hotels with international guests (Cichy & Wise, 1990). Ninemeier and Purdue (2008) added that RS menus are prime ways to cross sell hotel products and services by adding information about them in the body of the menu.

RS menus should share a theme with other outlets if possible and possess an attractive design that includes prominent display of the most profitable dishes. A specific menu unique to RS is the doorknob menu. All RS operations should have a doorknob menu for ease and facilitation during the busiest RS meal period, breakfast (Cichy & Wise, 1990). These menus are filled out by the guest the night prior and hung on the doorknob for pick-up and preparation for delivery at a specified time range, but some flexibility in time range (half-hour increments) should be applied.

2.17 Operating Trends

Popular press and trade journals in this area have generally focused on a specific component of the process, be it safety, differentiation, sanitation, or delivery methods. While this information was used in the coding process of analysis, it also served to generate a list of critical factors that were incorporated into a "panel of experts" phase of analysis. This assisted in ensuring that the resultant generic audit generated by this study includes any emergent principles and practices common to room service that may not be covered in the audits by hotel companies.

In the popular press and trade media, the literature on room service varied widely in terms of subject material. Stellin (2007) wrote of the need for hotels to redesign their room service menus, specifically to address customer trends and needs. Churchill (2006) discussed room service customer attitudes towards price and transparency of costs. Serlen (2003) related how room service can be adjusted to better accommodate the needs of the business traveler segment. A majority of these changes were focused on menu engineering and design. Serlen noted that more items were being offered to meet the changing consumption habits of the traveler, specifically in the areas of healthy foods, fast preparation and comfort foods. Serlen discovered that those who traveled 10-15 days a month ordered the most frequently from the RS menu and that a variety of balanced, healthier dishes were a predominant part of their orders.

Many of the works addressing RS have tended to focus on trends in that operation and are a snapshot in time of hotel strategies designed to accommodate the evolving needs of their guests, providing further insight as to policies that should be considered in the construction of an effective RS operation. As an example, for many years, some operations added a surcharge to the price to cover additional materials and resources, but there has been a trend in the past towards aligning RS prices with restaurant prices in the facility (Cichy and Wise, 1990). Turk (1979) noted that some facilities with multiple food and beverage outlets offered a multi-course meal experience pulling favorite or signature dishes from various outlets into one dining experience. Finn (1977) noted that some operations in larger facilities in urban settings dedicated specific service elevators that were set up to act as floating continental breakfast kitchens. Customer orders went directly to the elevator and the order taker prepped the meal in the elevator en route to the guest floor and delivered it directly and with surprising speed. At that point in time, Finn asserted that big heavy breakfasts were in less demand, and small continental breakfasts were increasingly popular. Many other operating trends were briefly noted by Cichy and Wise (1990), and included things such as guaranteed delivery times, supplemental menus that pull from outside area restaurants, and video ordering processes on the guest room television, to mention just a few. Ninemeier and Purdue (2008) stressed that RS could benefit from better oversight. They believed that managers need quality RS because a well managed operation provides a greater level of guest satisfaction, a more positive service image, and a distinct competitive advantage.

Bernstein (1989) related a list of typical room service "nightmares" (failures). Situations that included extremely late and unpalatable food, grossly incorrect orders, hallways and rooms strewn with days-old trays and cold food, and ridiculously high prices with hidden surcharges were just some of the cases the Bernstein noted as common in RS operations. Keen (1994), on the other hand, showcased the most unique and luxurious room service offerings in Hong Kong. Some examples included personal chefs to cook meals in the rooms, customized elevators with kitchens that are like mobile room service departments, personalized menu and nutritional planning assistance, and wine lists with over 1000 selections. Finally, an interesting piece by Hemp (2002) reviewed his first week as a room service waiter at a Ritz-Carlton Hotel and included a good overview of the training, set-up of the operation, and operating processes as seen by the worker.

Taken separately, these trade articles contribute little to the theoretical foundations of this study. However, in the context of an exploratory study such as this, every piece of information that can be uncovered, especially ones that depict operating processes can be valuable. The more information gathered from what is actually happening (or should happen) in room service operations in hotels, the more likely it will be that the process of identifying and coding the various practices will be robust and representative as an overview.

2.18 Service Quality and Quality Control

There are several different definitions of the term quality, and a vast amount of literature in the hospitality, marketing, and business disciplines that has focused on what quality is, and how it is developed, measured, perceived, and improved. For the purposes of this study, quality signified something of high grade, superiority, or excellence and focused on quality control, specifically in the delivery of products and services. Some background will help to contextualize the meaning of quality and quality control in this research. Charles Deming is considered by some to be the father of modern quality (Kandampully, 2007). He and a team of researchers went to Japan after World War II to assist in the rebuilding of the country's economy and improve the quality of goods manufactured there. Deming applied a technique of random sampling from the factory line to analyze the quality of the goods produced. His interpretation of quality was a complete absence of defects (Gabor, 1990). If the product created was free of any such defects, it was considered to be of high quality. Deming's work has been connected to Total Quality Management, a process where all aspects of an operation are designed, examined, and readjusted to eliminate all controllable defects that may arise as a result of the production process (Pfeffer, 1982). Although the process of random sampling ultimately developed into statistical quality control, the notion that a product free of variations from its intended design is considered to possess quality is of central importance.

Services, however, are not created on a factory line. Service can be defined as an act or deed in time that benefits the customer (Kandampully, 2007). Sometimes, there is a core product that can be measured, and other times a service's core product can be completely intangible. So, we cannot necessarily pull a service product from a factory line and examine it in the same way as Deming's procedure.

One of the distinguishing features of service is referred to as heterogeneity. Also known as variability, this characteristic is concerned with the occurrences of differences of variations in the outcome of the service product (Kandampully, 2007). People are involved directly in the production and delivery of service and service products, and therein is the potential problem. People are human, not machines, and make mistakes or alter their delivery of service from time to time, depending on environmental circumstances. A model that describes quality in the context of service production is needed and would be more appropriate for establishing an operationalized definition for this study. Gronroos (1984) developed a model of service quality that will be used in this study.

The Gronroos' Model of Perceived Service Quality examined the assessment of quality from the customer's perspective. In this model, quality is seen as a confirmation or disconfirmation by the consumer of the service in terms of how much difference there is between what customers expected and what they perceived they were given (Gronroos, 1984). Though the model is concerned with this gap between expectation and perception, it clearly laid out the components of a service product. The model describes a service as being a combination of technical outcome and functional outcome. The technical outcome is the "what," the core product, while the functional outcome is the "how," the manner in which the product is delivered. Both components are considered to be equally important in the successful delivery of a service. This study was not as concerned with other elements on the consumer side of the model, such as customer needs, learning, and word-of-mouth influences. The production side, however, which demonstrates how functional and technical outcomes combine to create a perceived image of the service, is precisely how this study viewed service production and quality.

There were many studies that discussed quality and quality perceptions, most often from the consumer side. While some of the other studies may have defined quality as well, none seem as salient as Gronroos' model, even though other studies have supported his model and in their results connected service quality to satisfaction. Augustyn and Ho (1998) and Weber (1997) conducted separate service quality studies that indicated perceived quality on the part of the guest leads to higher levels of satisfaction and subsequent repeat patronage intention, an important outcome for lodging operations. These studies demonstrated that guest perception of service quality provided benefits to operators who make pursuit of quality a business imperative. As this study is focused on the creation of a viable instrument to measure an operating process, practical significance can be achieved if such proposed measurements are shown to provide a benefit or return on investment to operators that choose to implement them.

Having defined and discussed service and service quality, this study examined the concept of quality control. If service quality is of interest to the provider, but is difficult to measure or homogenize, how has previous research advised its control? The answer is some form of quality assurance or quality control program for the operation or process in question.

Quality control programs, also referred to in hotel operations as quality assurance, are focused on the concepts of standards and expectations (King, 1984), one being the prevention of deficiencies, not remedies of them. According to King, quality programs in an operation were concerned with standards, products and services, production methods, systems, and other specific processes. Standards, products, and procedures should be developed to meet consumer's needs at the lowest possible cost. Other specific processes were those considered to be any procedures in the operation that were aimed at minimizing the failure to meet service standards, referred to as "non-conforming output." The system King defined was especially relevant to this work, as it is needed to ensure that operational output met consistent standards. Another key factor was a comprehensive and effective service-oriented quality control system that required some sort of process that compares the real output to the ideal output. This is in alignment with the foundational Gronroos' model.

Though King's work is substantive and directly linked to the thought process behind this review, many others have written about quality control programs in hospitality. Comen (1989) discussed how such programs have been inconsistently designed or executed in hospitality. His work provided several salient guidelines for creating an effective, long-term program that will prevent, halt, or reverse a decline in perceived service quality. Comen's (1989) and King's (1984) studies provided a link between quality as a concept and the pursuit of quality in an actual operation. Karapidis, Athanassiadis, Aggelopoulos, and Giompliakis (2009), Glover, Morrison, and Briggs (1984), and Pei, Akbar, and Fie (2006), have all contributed to the area of quality control research with specific unit-level studies of hospitality operations. Narangajavana and Hu (2008) investigated hotels in Thailand when the government developed a new ratings system. These studies focused on the effectiveness of quality control systems, such as rating systems and standards. Findings in these studies supported the idea that quality programs in hospitality operations produced improved satisfaction or perceptions of satisfaction amongst its guests. They also reviewed potential obstructions to their use or adoption. More importantly however, each suggested, as did King (1984) and Comen (1989), ways to improve quality programs through precise design, consistent application, and objective measurement.

Mohsin (2007) characterized service quality in research as an area of extreme debate and diverse findings. As competition increased drastically in the lodging industry in the 1990s, more hotels began to focus on service quality as a product differentiator. Mohsin felt that operators must become aware of how improvements to service quality, such as prompt food delivery to a guest room, positively impacted the perception of the operation. In his study, Mohsin (2007) indicated that service quality's largest issues for a hotel service offering included the way management perceived customer importance, lack of training, staff behavior and attitude towards service provision, and internal communication issues. Mohsin (2007) also identified RS critical factors for perceived service quality: prompt response from the order taker; variety in the menu; prompt service and delivery; food/beverage quality; and perceived value for the money spent.

In these quality studies and in other literature discussing service quality in general, it has been proposed that measurement of the service processes against their intended outcome is an important part of any quality program. Few of these works discussed how these measurements should be made or through what instrument they should be conducted. This helped connect the literature review to the next area of interest: the effort to continually improve service through the application of operating audits.

2.19 Operating Audits

Having established a reasonable understanding of what quality is in services, as well as the need to carefully approach and measure it in a service operation; this study investigated what the best way to accomplish this is. The goal of this research was to create an instrument that would help an operator compare the actual outcome to the intended or ideal outcome. The resulting instrument would be used as an operational audit for RS.

There were several seminal books on operating audits. Reider (1994), Thornhill (1981), and Flesher and Stewart (1982) have all written comprehensive overviews of the audit process. In these pieces, they defined auditing, identified its benefits, and advised on the design and implementation process for auditing efforts.

Although several works discussed both, operating audits differ from internal audits. Reider (1994) noted that there was no one commonly recognized definition for the term, and listed 10 commonly accepted definitions. Thornhill (1981) described operational auditing as the application of internal auditing techniques without the typical financial controls. Flesher & Stewart (1982) agreed, and added that an operating audit was the examination of activities that were non-financial in nature. It is clear that operating audits evolved from the financial process of internal auditing. This study is focused on developing an audit that reflects best principles and practices currently in use in the hotel room service operations. This dissertation utilized Reider's (1994) fourth description of auditing as a control technique for evaluating the effectiveness of operating procedures.

Flesher and Stewart (1982), Reider (1994), and Thornhill (1981) linked operational auditing to the model of the "Three Es": economy, efficiency, and effectiveness. In this model, economy was seen as the cost of operation, while efficiency was the method of operation, and effectiveness was the operating result. While this study was not initially concerned with the economy portion of the 3-E's model, it was directly concerned with the method and results of operating efforts, or the other two Es. This model provided an excellent linkage to the previous section on quality control, where the need for a method to measure these operating components was set forth.

Each author has developed a list of benefits or reasons why an organization should perform audits. Reider (1994) felt the primary benefit of audits was to identify problem areas or areas requiring improvement. Thornhill (1981) and Flesher and Siewart (1982) agreed that the primary benefits of auditing were involved with identifying opportunities to increase revenue or reduce expenses. As an observation, the difference of more than ten years in Reider (1994) and the other two author's works indicated how operating audits have evolved from their financial origins. The latter two's works seemed to return often to the concept of cost savings and other financial goals, while Reider's works mentioned them as well, but not as prominently. This is not to say that any business should not have the goal of overall cost efficiency in its operation; however, the main focus of this study was on operating performance, not financial.

Other areas of concern within this topic included the planning phase of the audit process. Reider's (1994) approach was to identify information that should be obtained, select appropriate sources of information, and identify critical areas and key activities to audit. The methodology of this study followed that format. Reider's approach appeared to make the assumption that the organizations generating these audits have indeed identified what information should be measured. As a preview to the methodology section of the current study, a panel of experts assisted in ensuring that any additional components or classifications of room service procedures that might be considered critical or key areas that were not present be added.

Reider (1994) described an operating audit as a tool for the assessment of the effectiveness of operating procedures. Thornhill (1981) and Flesher and Siewart (1982) established processes for conducting the audit, analyzing the results, and preparing the final report. In many cases, in addition to a physical inspection of the operation, a review of pertinent historical documents at the workplace setting was conducted. The analysis of RS operating audits will be discussed in the methodology section.

2.20 Hospitality Operational Audits

There were some specific works on hospitality and operational auditing. Willborn (1986) wrote on quality assurance audits in hotels. His work provided a general outline of how to approach the design and execution of the process and often mirrored the processes as described by Reider (1994). There have also been a few key works on the development of operating audits for specific hotel departments by Moreo and Savage (1990), Moreo et al. (1997), and Wood et al. (2005). Each of these works focused on a different operating department within a lodging facility, although RS was not included.

Willborn (1986) identified the quality assurance and audit process for a hotel as a cycle with five steps. As with Gronroos' model, quality was seen as a state of perception on behalf of both the provider and the consumer, and the cycle required the establishment as well as an assessment of what these perceptions were, or should have been, in order to best evaluate and manage them. The first step in the cycle was for the hotel's management to establish what their quality image was. The hotel's marketing and positioning efforts to distinguish itself from or as a part of a particular tier or market segment created this image. The next step, as with Gronroos' model, was the establishment of the quality expectations of the consumers. These were developed by prior experience, word-of-mouth, and the positioning efforts in the first step.

The third step was concerned with the guest's quality perception. In other words, how close or far did the operator come from fulfilling the expectations of the guest? Willborn (1986) noted, like Gronroos' (1984) interpretation, but with more emphasis on the relationship, that the second and third steps of the cycle are greatly related to the first. In other words, the quality image is not just how the hotel depicted itself in strategic media plans, but also how well or how close the actual operation came to those image promises. These three steps led to the development of a quality assurance program in step four, which was supposed to reflect the procedures necessary to create the image of the facility. These are often referred to as standards or operating policies in the hotel industry. It is these written standards, explaining, describing, or directing processes within an operation that were the source of the content analysis portion of this study. The audit itself, Willborn's step five, is developed from the procedures of step four, and used as an independent test or assessment of how well an operation is in compliance with the procedures. Willborn stressed that hotel management needs the independent evaluations of operations, specifically due to the valuable information that they provide operators who may not be able to inspect all aspects of a given process. He added that, for such audits to be of value, they must be thorough, comprehensive, well conceived, and consistently executed.

Moreo et al. (1990) felt that hospitality operational auditing could be improved by studies such as he and his colleagues conducted, primarily because lodging companies tend to keep such information confidential and proprietary, and as a result, there was very little information shared that would help to document current practices and see where industry trends or emphasis might be currently focused. This study, focused on hotel accounting and guest processing and relations' practices at the front office, suggested that audits for a hotel should be comprehensive, yet flexible enough to be custom-tailored to fit the needs or goals of the operation. It noted that all employees, management, or staff should be involved in the audit process, making it a self-audit tool of sorts, whenever possible. Finally, for the audit to be effective, it must have a follow-up phase that directs action and thought towards where deficiencies are and how they can best be corrected.

Wood et al. (2005) conducted a similar study of hotel housekeeping operational audits. In addition to utilizing content analysis to develop an over-arching view of all of the components of a housekeeping operation, this study developed a summary list of the most important elements in housekeeping, as demonstrated by the frequency with which certain operational aspects or procedures appeared in the various subject audits of the study. This study also utilized expert panel and field testing, which will be replicated to a degree in this dissertation. Wood et al. (2005) also supported the use of the audit as a training tool in addition to it being a tool for assessment.

2.21. Conclusion

This literature review was focused on setting a foundation for this study on RS principles and practices as they pertained to operational audits. The literature reviewed discussed the established, chronological process behind room service, utilizing trade articles and academic textbooks. This information served as a guideline and a framework for the coding process of the content analysis of historical RS documents. The review also discussed foundational theory in service quality, supporting the concept that giving the customers what they expect will provide a return on the business' investments. Finally, a review of the literature and research in the areas of operational and hospitality auditing provided the link between the goals of service quality and customer service, and a viable manner of measuring how close or far an operation is from achieving that perception. In other words, hospitality operating audits were viewed as a necessary tool for fine tuning a business to be appropriate and efficient, and given the potential return in RS, coupled with the inconsistent manner in which it is currently seen or practiced by guests and staff, the practical and academic significance of this study are in alignment with the needs of the industry. A review of the literature has provided the foundation to design this study, and the support to conduct it. In the next chapter, the methodological design and execution of the research are discussed.

PART III

3. METHODOLOGY

The main objective of this study was to identify the best practices and principles for in-room guest dining that currently exist in first-class and luxury hotels. This was an exploratory study because there has been little research of any kind completed on best principles and practices in RS in first-class and luxury hotels. There were three motives for conducting an exploratory study of this nature. An exploratory study focuses on the investigation of little known or understood phenomena, identifies important and specific meanings, and generates direction for future research (Marshall & Rossman, 1999). This dissertation attempted to collect as much historical data as possible, with the intention of gaining insight into the current work practices, standards, and processes of RS operations in first-class and luxury hotels.

This dissertation used a mixed-methods approach as its methodology as defined by Johnson and Onwuegbuzie (2004) as a type of research where the researcher combines quantitative and qualitative research techniques, methods, approaches, concepts, or language into a single study. A key feature of mixed-methods research is its methodological pluralism, or eclecticism (Johnson & Onwuegbuzie, 2004). This pluralism can frequently result in superior research when compared to monomethod research. If exploratory research is to be accepted as necessary and foundational to research in the same area of investigation, it must be robust and comprehensive. Mixed methods can be especially useful in a study such as this one, where little research has been previously conducted.

There is an ongoing discussion in academia concerning the efficacy of qualitative versus quantitative research. The goal of mixed-methods research is not to replace either of these approaches but rather to draw from the strengths and minimize the weaknesses of both in a single research study (Marshall & Rossman, 1999). Johnson and Turner (2003) identified and defined the fundamental principle of mixed research. In a mixed-method approach, researchers should always attempt to collect data from multiple sources, using different approaches, and in such a way that the resulting combination is likely to generate complementary strengths and non-overlapping weaknesses. Brewer and Hunter (1989) stressed that effective use of this principle is a major source of justification for mixed-methods research because the analysis and outcome will be superior to monomethod studies.

There are many characteristics in mixed-methods research that made it appealing to adopt for this study. Johnson and Onwuegbuzie (2004) noted that words, pictures, and narrative can be used to add meaning to numbers, and, that conversely, numbers can be used to add precision to words, pictures, and narratives. The researcher that uses mixed methods can generate and test a grounded theory, often with the ability to answer a broader and more complete range of research questions because there is no confinement to a single method or approach. Brewer and Hunter (1989) added that a researcher could use the strengths of an additional method to overcome the weaknesses in another method by using both in a research study. This in turn will provide much stronger evidence for a conclusion or a more complete overview through corroboration of findings. Finally, Johnson and Turner (2003) stress that mixed-methods research can add insights that might be missed when only a single method is used, contributing to the effective generalizability of the results. Overall, qualitative and quantitative research used together produces more complete knowledge, necessary to inform theory and practice. For these reasons, and since this was an exploratory study that sought to discover what first-class and luxury hotels are doing in their RS operations, mixed methods seemed appropriate to adopt.

3.1 Research Overview

The data collection and analysis in this mixed-methods approach occurred over several stages. Each of these stages will be discussed in further detail later in this chapter, but it is prudent at this point to review the process concisely prior to proceeding. The first stage was the content analysis stage. In this stage, historical work documents, including but not limited to operation checklists, policy memos, and training manuals, were gathered from various hospitality entities that own or operate luxury, lodging facilities that offer RS to their guests. Manual and computer content analysis were conducted in order to generate a comprehensive listing of all of the themes that emerged from the documents. In the next stage, the initial listing of themes, a first draft of the generic audit that was the intended goal of this study, was presented to a panel of experts for review and commentary. The panel consisted of hospitality executives and employees that are directly involved with RS operations. A convenience sample determined the selection of the panel, and members were added through any referrals, prior relationships, or industry contacts that were willing, able, and eligible to participate. The panel assisted in verifying the completeness of the list. They also participated in the development of the final coding schema. Through this process, the panel helped to lend face validity to this study. In other words, an instrument can be said to have face validity if it "looks like" it is going to measure what it is supposed to measure (Merriam, 2002). By utilizing industry professionals to support the inclusion of components in the audit, the study verified that it is measuring what it intended to, the identification of key operating procedures in RS. After the content analysis and panel of expert review was conducted, revisions were made to the list of themes, and the format was revised to create the next draft of a generic audit. At this point, the resultant audit document was made into an operational audit that was concise, sequential, and easy to understand and utilize.

The next stage was to field test the audit in one actual first-class/luxury hotel RS setting. Several luxury hotels with which the researcher has contacts were solicited to be a test hotel for the audit, and the first willing hotel was utilized. After field testing, further revisions were made to the audit, and it was returned for a final review to the panel of experts. A final draft of the audit was then completed. That draft was condensed into a survey instrument by the addition of a Likert scale rating the relative importance of each line item in the audit template. The scale is a summated rating scale, usually with five to seven points, used for measuring attitudes (Marshall & Rossman, 1999). It was a method of ascribing quantitative value to qualitative data, to make it amenable to statistical analysis. A numerical value was assigned to each potential choice and a mean figure for all the responses was computed at the end of the evaluation (Johnson & Turner, 2003).

This instrument was then posted online on an Internet survey portal for the next stage, which consisted of inviting RS professionals at any level to rate the line items on the audit. In addition to the audit itself, there were a few short qualifying and demographics questions for each respondent. It was during this stage that the research revealed differences among groups of respondents with respect to the importance of certain line items or coded themes. After these stages were completed, the study was able to address the initial research questions stated in the introduction of this study. This multi-stage, mixed-method approach delivered enough data to support the content of our generic audit, and the quantitative testing added insight into variances among property types in the execution of RS operations in first-class and luxury hotels.

3.2 Content Analysis

Berg (2001) described content analysis as a process for systematic inferences about meanings in communication that objectively identifies special characteristics of messages. This study utilized historical documents for analysis. Specifically, this study defined historical documents as any work-related checklists, audits, inspections, memos, training manuals, policy manuals, or standards manuals related to the operation of the RS department. While qualitative analysis does not necessarily concern itself with generalizability, it is interested in representativeness (Merriam, 2002). In other words, sample size in qualitative analysis is not dependant on a specific size. Rather, a sample size is considered adequate in qualitative analysis when coding schemes begin to repeat themselves and no new themes emerge during the initial coding process. Generally speaking, a sample size of seven to nine should be large enough to generate a robust coding scheme (Keppel & Wickens, 2004). This study collected an array of historical documents from seven different hospitality entities, and the population utilized for this sampling includes any first-class/luxury hotel or hotel company with RS operations. This included any independent or franchised operations, either at the unit or the corporate level, which held, created, or utilized documents describing the requirements, standards, and processes in a RS operation.

An important part of the analysis portion is to determine the unit of measurement. For historical documents, the most relevant units of measurement are words, phrases, and themes (Kassarijian, 1977). As recommended by Mehmetoglu and Altinay (2005), historical documents from each organization should be first analyzed separately to identify emerging themes within particular organizations. Themes that emerge from each document are then compared across organizations in order to identify common practices. Such a multiple analysis approach can lead to a more sophisticated understanding of the phenomena being investigated (Mehmetoglu and Altinay, 2005). This dissertation was designed to present as much information as possible to assist with future research efforts.

Content analysis was completed using Atlas/ti, a computer-based program for users to manage, organize, and support research using qualitative data (Muhr, 2005). A hand-coding process was utilized first, prior to the computer analysis. Hand -coding is a necessary component of preparing the historical documents for analysis in Atlas/ti. The coding process was simple and natural. The unit of measurement must first be defined. Berg (2001) identified "theme" as a useful unit to count, defining it as a simple string of words, or a sentence. Six initial dimensions for the construction of the coding scheme were developed from the literature. Cichy & Wise (1990), Ninemeier and Purdue (2008), and Rutherford (1990) all discussed RS as a chronological set of processes that follow a general sequence. As established in the literature review, these six areas, which were utilized as super codes, include pre-preparation, order taking, order routing, order preparation, order delivery, and post-order activities. These six themes developed from the literature can then be considered as major themes utilized in the process of coding (Berg, 2001). After identification of passages in these historical documents, further word, concept and theme coding was conducted. Specific codes were developed that reflect the descriptions and requirements of all of the passages revealed in the data. All codes were developed using the natural text of the passages. The analysis section of this work will relate the final coding schemes developed from the data using both text and data. Current computer programs are not sophisticated enough to identify emergent codes without being directed to do so from a list of pre-generated options. Identified codes were entered into Atlas/ti in a list developed during the process of manual examination.

Muhr (2005) described the process of coding data for analysis by Atlas/ti. First, all of the data (historical documents) gathered are entered and associated with the program in what is referred to as a hermeneutic unit (HU). This creates a single file out of all the various data components, allowing easy access to all the sources in one place. Text passages are then assigned established codes previously determined and identified by the researcher. These passages tie together lines of text, referred to by the program as quotes, to certain codes from the established list. Upon completion of initial coding, the program is then able to look at patterns across all documents, and generate an assessment of codes fully or partially shared by multiple organizations. With Atlas/ti, the researcher can access specific features that can assist in making connections between various coded concepts, especially in terms of distinct forms of useful diagrams and other visual output (Muhr, 2005). From this analysis, the researcher can build semantic, propositional, and terminological networks that represent relationships between the content of the varying documents and organizations. These results can then be further analyzed by the program's identification of coding that is prevalent in all data sources.

The researcher guided this process by tacking memos of interest to multiple quotes from different documents, serving to further organize the data in a meaningful way. Output from the program was generated via a complied report that groups all codes, super codes, family codes, memos, and networks into one amalgamated file (Muhr, 2005). Additionally, Atlas/ti also assists in producing graphic representations of the existence, relationship, frequency, and strength of themes across data sources. This research produced several such outcomes, which will be presented in the analysis and discussion sections.

3.3 Panel of Experts

After a list of codes based on historical RS documents was developed, a draft of the generic audit was compiled. This list, with all of the varying codes developed during the content analysis stage, was presented to a panel of experts for review for face and content validity. Yousuf (2007) describes the panel of experts as a group process involving an interaction between the researcher and a group of identified experts on a specific topic. This technique provides different opportunities to researchers than does the typical survey/intercept research. According to Yousuf (2007), the essential components of the panel of experts included the communication process, a group of experts, and essential feedback. Linstone (1978) added that best applications of such panels are whenever policies, plans, or ideas have to be based on informed judgment. This technique is useful where the opinions of practitioners are needed but time, distance, and other factors make it unlikely for the panel to work together in the same physical location. Linstone (1978) noted that the best applications for this technique included, but are not limited to gathering current and historical data not accurately known or available, putting together informative and accurate models, and delineating the pros and cons associated with policy and standard options. All of these benefits directly addressed the goal of this study, which was to make a comprehensive generic operational audit that was realistic, appropriate and useful to the hospitality industry.

Due to the potentially wide variance in standards for RS among organizations, a panel of experts was adopted. This panel consisted of hospitality field experts with direct involvement or oversight of RS operations in first-class and luxury hotels. Morgan (1997) asserts that such a panel, consisting of five to seven members, will produce optimal output with minimal conflict. The panel of experts approach utilized a systematic process of information gathering. The technique is most useful where the opinions of experts and practitioners are necessary (Morgan, 1997). The use of such a panel is a form of interrater reliability that addresses the consistency of the implementation of a rating system (Yousuf, 2007). It determines the extent to which two or more raters obtain the same result when using the same instrument to measure a concept (Berg, 2001). Though this study utilized more than one rater, the panel further strengthened the process through independent verification of the coding schemas.

Brooks (1979) outlined the procedure for effective use of a panel of experts, and this study followed that outline. First, the requirements for a panel must be decided, and willing participants that meet these standards must be identified. In this study, the next step involved the gathering of specific information related to the research questions and the topics being reviewed by the panel. This information was presented to the panel for its review. This "gathered" data source was the first draft of the generic audit, created through the content analysis of historical lodging documents related to RS operations. Specifically, the researcher sought feedback on the completeness of the content for the audit, as well as the format of the audit as it pertained to its ease of use. Feedback from the panel was then organized and redistributed to the entire panel. In this study, that redistribution took the form of a revised audit draft. The panel again reviewed the revised document along with panel feedback notes. In the first review, the panel looked solely at the content and used its personal expertise and current workplace standards to compare to the audit. After the initial analysis, panel members were expected to review the document with a focus on comparing their responses to that of the group as a whole. In cases where individual responses varied significantly from that of the group, the individual was asked to provide a rationale for differing viewpoints (Easterby-Smith, Thorpe, & Lowe, 2004). It is in these two phases that the audit took shape, ultimately reconciling standards established in the literature with actual RS practices that are in place in first-class and luxury hotels today.

The panel of experts had several advantages appropriate for a mixed-methods study (Morgan, 1997). It was especially useful for this study as the problem of creating the optimal generic audit for RS does not lend itself to precise analytical techniques, but rather can benefit from subjective judgments on a collective basis. Additionally, the participants, by purposive selection, represented diverse backgrounds with respect to expertise. It is a viable approach to use when circumstances make frequent, in-person group meetings infeasible. There were other advantages to this technique, including the prevention of groupthink by dominant personalities, simplicity and ability for the respondents to express themselves in a non-constricted way with regard to format, response time, length, or nature (Seidman, 2006). Morgan (1997) adds that experts tend to present widely different opinions based on their personal experiences, making for a rich amount of information arising during the reviews. This can provoke better collective responses than a single person might, be it in communication or in writing. In other words, the panel's interactions sparked further insight and higher quality responses. In this study, the panel was conducted through an open conference call monitored and directed by two researchers in order to promote and facilitate the highest possible interaction.

3.4 Field Testing

Marshall and Rossman (1999) define a field test as the act of testing a device or method under actual operating conditions. Such field testing generally encompasses development, testing, and evaluation of the method and its aim in the case of this study was to contribute to the generalizable knowledge on RS operations. It is important that such testing is conducted under actual working conditions with its intended use in order for it to be considered a valid test (Keppel & Wickens, 2004). Merriam (2002) noted that there are three types of field testing: direct observation, participant observation, and qualitative interviews. This study utilized the direct observation approach for the field testing of its operational audit in a RS setting.

Field testing is different from laboratory testing or developmental testing (Merriam, 2002). Laboratory conditions are often strictly controlled in order to best isolate variables and create a rigorous design. Field testing can often reveal design flaws or structural issues with the test subject that were not considered or went unnoticed during previous stages of analysis. Field testing can be useful for spotting a wide range of interaction problems that would not necessarily appear in a clinical setting (Keppel & Wickens, 2004). In a hotel, these can include problems or issues such as heavy demand, operational crises with operating equipment, varying degrees of quality of employees and employee training affecting the production or delivery of food, and others. It is so hard to predict these kinds of interactions, even when heavily considered by knowledgeable experts, that it is not possible to foresee all of the varying situations that may have an impact on the outcome of the service provided. Therefore, field testing helped to ensure the usability of the resultant audit create by this study. Any noteworthy revisions in the processes during field testing were identified, along with any additional commentary from field researchers on the efficacy and usability of the instrument.

3.5 Quantitative Analysis

The next step in the research process for this study was to conduct a survey of hospitality personnel directly involved with RS functions in a luxury lodging facility. A condensation of the generic audit, developed through content analysis, supported by input from a panel of experts, and field tested in a first-class/luxury hotel, served as the survey instrument for this study. The panel of experts reviewed the final version of the fieldtested generic audit and rated the most important elements using a Likert rating scheme. The highest rated of the coded items on the generic audit were presented in an online survey format. Hospitality professionals with RS experience were invited via e-mail to review the identified operational processes and rate them, using a 5-point Likert Scale. The responses from participants were then analyzed.

One of the research questions developed for this study was, "What are the noteworthy, differences between these standards across company or asset types?" Based on this research question, an appropriate methodology for investigation of this query was multiple analyses of variance (MANOVA). MANOVA is used to determine the main and interaction effects of one or more categorical (non-metric) variables on multiple dependent interval (metric) variables (Hair, Black, Babin, Anderson, & Tatham, 2006). MANOVA uses one or more categorical independent variables as predictors, with more than one metric dependent variable, providing insights into the nature and predictive power of the independent measures, as well as the interrelationships and differences in the multiple dependent measures (Hair et al., 2006).

With the data from this study, the focus was on how the dependent variables, which were the various operating processes and procedural standards in an RS department, interacted with independent metric variables. Generally, in exploratory research, there are many potential independent variables (Locke, 2003). One of the research questions in this study was, "Can we create an overview of these identified practices that will address all, if not most types of first-class and luxury hotels?" If the overview, created in the form of an operational audit, is to best serve its function as a template for any operation to use, significant differences between operation types should be identified. This approach assisted in the audit template being more appropriate for use in a wider range of hotels.

Independent variables selected to support the objective of creating a wide, generic template for application in the field included hotel size and hotel type. Size was categorized as small (under 200 rooms), or large (200 rooms and over). Hotel type was categorized as resort and non-resort. The presence of a free-standing RS department versus a RS function that operates out of an existing outlet's facility within the hotel is another variable that was examined. Finally, the hotel's operation as a franchised property as opposed to an independent facility was investigated. Therefore, the quantitative research section focused on observing the type, size, and franchise status of the hotel, as well as the operating structure of RS within the hotel where respondents work, and their perceptions about the importance of the different operational processes identified in the qualitative analysis stage.

Using these independent variables, four hypotheses were developed: H1: There is no difference in the perceived importance of key room service practices between employees of a large hotel and employees of a small hotel. H2: There is no difference in the perceived importance of key room service practices between employees of a resort hotel and employees of a non-resort hotel. H3: There is no difference in the perceived importance of key room service practices between employees of a hotel with a freestanding room service department and employees of a hotel without a freestanding room service department. H4: There is no difference in the perceived importance of key room service practices between employees of a franchise hotel and the employees of an independent hotel.

The research questions previously noted served as a guideline for the creation of the survey instrument and the selection of an appropriate methodological tool for analysis. Although, analysis of variance does not require the establishment of a hypothesis per se (Hair et al., 2006), this study was aimed at not only identifying current practices in RS, but also in designing an audit that considered variation in processes across different operations. These hypotheses and MANOVA strengthened the overall analysis of this study, and created opportunities for further studies in the future.

3.6 Conclusion

In summary, this study was a replication and an extension of previous research on hospitality operational audits (Moreo & Savage, 1990; Moreo at al., 1997; Moreo et al., 1999; Wood et al., 2005). It was a replication in the sense that it followed the isolation of a specific set of functions in an operating department, and then utilized historical corporate documents to identify the general and specific processes and criteria that composed the standard or the standard's desired outcome. It was an extension in the sense that several additional steps were added to the methodological process used by the earlier researchers. The extension in this study included the addition of a panel of experts to assist in verification of audit components, a field test of the resultant audit at a lodging facility, and a mixed-method application to quantitatively explore differences in perceived levels of importance among groups on specific operating policies. Using the methods outlined in this chapter, this dissertation analyzed the data from each phase in the analysis section.

PART IV

4. RESULTS

4.1 Content

Analysis A convenience sampling approach was utilized to procure the necessary historical work documents from lodging organizations. The researcher contacted organizations that employed personal or professional contacts in order to request access to their RS policies and procedures. Academic colleagues were asked to reach out to their industry contacts as well. The sampling process for these lodging companies was also purposive. After reviewing all the organizations willing to share these documents, selections were made to include a wide array of company types and structures. This was done in order to get the largest and broadest scope of RS policies currently in place in the lodging industry.

Eight hotel organizations were selected as participants. Two of the participants were independent resort facilities, one smaller facility located in the northeast United States with less than 200 guestrooms, and one large casino resort located in the western United States, with more than 2,000 guestrooms. Three of the participants were hotel franchisors. Three of the participants were hotel management companies. The franchisors and the management companies had multiple holdings all across the U.S., varying in size from below 100 guestrooms to over 500 guestrooms. All participants operated upscale, full-service first-class/luxury hotels, the specific target segment for this study. This broad sample provided a rich cross-section of operating policies and procedures as well as specific and unique approaches to the in-room dining experience.

Participants provided electronic and hard copies of the requested data to the researcher. The documents were reviewed to identify sections that were specifically about room service operations. Data collection generated 1,163 pages of specific RS related documents. These documents came in many forms. Documents included corporate-level operating audits, training manuals, policy manuals, orientation manuals, and memorandums. All selected documents described or related procedures that had to do with the administration, preparation, or execution of the in-room guest dining experience.

Content analysis was performed on these documents. Because one of the goals of this study was to generate a comprehensive generic audit that could be utilized as a form of self-assessment, an initial set of super codes was generated from existing literature. It was decided to utilize a set of super codes that categorized the activities being coded in terms of when they occur in the production process as seen in Table 1. The added benefit of coding in such a sequential fashion is that it keeps the data organized and focused towards the study's goal of creating a tool that is easy to understand and simple to utilize by any of the personnel in the RS department. The super codes were extracted from the literature (Cichy & Wise, 1999; Davis & Stone, 1985, Ninemeier & Purdue, 2008; and Rutherford, 1990), and consisted of six initial categories; Pre-Preparation Activities (PPA), Order Taking (OT), Routing (R), Order Preparation (OP), Order Delivery (OD), and Post-Order Activities (POA). To this list, two additional super codes were added by the researcher after precursory reviews of the data indicated that there were several items that could not be readily categorized into the existing, literature based codes. The two additional super codes were Administration (A), and Other (O). Operationalized definitions were developed for these super codes in order to assist in the consistent coding of all items. These eight coding definitions are included in Table 1.

Table 1. Super Codes

PPA	Pre-Preparation Activities Definition: Any activities completed in order to prepare for a given shift. These include duties such as inventory and stocking, cleaning, and processing opening paperwork.
OT	Order Taking Definition: The process of taking an order for a RS delivery. This may be done through employee-guest interaction over the telephone, or through a written process, such as a doorknob menu. Often, the process, where possible, includes some degree of suggestive selling to the guest.
RT	Routing Definition: The process of transmitting an order to the appropriate production facilities in the hotel. This can include separating portions of the order to be completed and sending them to the kitchen or bar areas of food and beverage production. In many cases, this process is completed electronically through existing communication infrastructure.
OP	Order Preparation Definition: Any activities directly related to the production of food and beverage and the preparation of appropriate equipment for the delivery of that order. This can include preparation done by the kitchen or bar staff, or by the delivery personnel.
OD	Order Delivery Definition: The act of delivery of the finished order to the guestroom. This may include the set up of eating tables or trays and the presentation of the order to the guest.
POA	Post Order Activities Definition: Activities conducted after an order delivery has been made. This includes the processing of the guest payment, as well any other activities that are completed at the end of a shift, such as cleaning, stocking, and processing closing paperwork.
ADM	Administration Definition: Duties of a managerial or supervisory nature in the RS department. This includes inspection, employee management, menu engineering, policy development, and other related duties that are not specific to a particular chronological portion of the RS process.
O	Other Definition: All other activities that cannot accurately fall into one of the other existing super codes. Many of these may be organization-specific or hotel-specific.

After the establishment of the super codes, the coding process was conducted. The initial round of coding was completed directly on the documents, in the margins of the page, highlighting passages and making notes that were focused on the accurate and appropriate breakdown of each line item from the historical documents into its simplest elements. An open coding approach was used. Open coding utilizes specific words and passages from the text that help to describe the action naturally in as broad of a sense as possible. Wherever possible, exact and existing words were extracted from the text and used to name the code. This coding process creates tiers of codes, with each passage belonging to a super code category, and up to three or four more codes that further identify the passage's general intent without specificities. In general, most coded passages were limited to four tiers of codes or less in order to keep the coding process manageable.

Coded passages were entered as hermeneutic units into the content analysis software, Atlas/ti. As each passage was entered into the program, the super codes and any other initial tiered codes were entered along with the passage. The initial coding process took several weeks, and produced 665 coded passages, containing 248 unique codes. Most codes were designed to be a shortened abbreviation of the word that best described the activity being conducted, for ease of identification without a legend. In many cases, coded items could be used in any tier, depending on the context of the passages. A complete list of the initial codes by name is included in the Tiered Coding List (Appendix A), along with short definitions of each code where needed. Codes are separated by tier, and all codes are identified in the highest tier in which they were utilized, though they might appear in more than one tier. Only the super codes were limited to the first tier, as they were general headers that were used primarily for organization as opposed to identification.

In addition to the researcher's coding efforts, a second individual was tasked with coding the passages in Atlas/ti separately from the researcher. The second rater was an academic with high levels of familiarity with open coding and content analysis processes. Both coders worked on the data separately and without communication until both had completed a full coding of all 665 passages. Though the super codes and the tiered codes were shared between the coders, coding was not limited to existing codes established by the primary researcher. Both coders used existing codes when they were seen as accurate and appropriate. At several points in the coding process, both coders had to add additional codes they felt were more accurate in deconstructing the major themes in the data. Those codes were defined and added to the overall coding list in Appendix A. More than one coder was used in order to ensure inter-rater reliability, which, in turn, generates higher levels of face validity for the results.

After the separate coding processes were completed, the coders met on several occasions. Initially, they met in order to compare and discuss differences in their coding schema and application. The coders met on three following occasions in order to reconcile differences in code selection, context, and clarity. An additional goal for the two coders was to collapse as many coded passages into more general coding schemes in order to reduce the number of line items generated. Since one of the primary goals of the study was to produce a generic audit that could be used as a self-assessment tool, it was determined that more than 200 line items would not be manageable in terms of the size of the document and the amount of time required to complete such an audit.

Collapsing and condensing coded items took several weeks to complete. The initial approach included the identification of codes that represented specific actions that were similar in nature. For example, passages regarding the retrieval of guest trays or tables after consumption often used codes extracted directly from the line item. Depending on the source and the writing style of the document, the activity of tray retrieval was described using different descriptions. Tray retrieval was initially coded using several different codes: retrieve trays; scan for trays; inspect floors for trays; and walk the hall gathering trays. Circumstances such as this were prevalent across several areas and in several super code categories.

The two coders met and discussed each of these sets of similar items in detail, reviewing when the task was to occur and what the specific action was to be completed. When both coders were satisfied that differing codes were representing similar tasks, they collapsed the coding into one homogenous code. In the example of tray retrieval, it was determined that retrieval (RETR) would be a code that would satisfy the spirit of all the differing codes previously used to identify the passage. If the tasks identified were to be completed at a different time during the RS process as identified by its super code, the codes were generally not collapsed. The reasoning behind this was that a 24-hour room service operation was very likely to have more than one employee performing the duties during the course of the day, and the audit could become convoluted if it was not clear as to when precisely the task was to be performed in terms of the natural sequence of events in RS.

Whenever there was a difference in coding that could not be collapsed with the agreement of both coders, a third coder was contacted. The third coder was an industry professional with over 20 years of experience in the first-class and luxury hotel food and beverage segment. The third coder was asked to use the existing coding schema to breakdown the passages in question and provide insight. This coding was free of any commentary or influence from the first two coders. The third coder's information was then shared by the first two coders, and a determination was made how and if the items should be further condensed. An example of the insight provided by the third coder would be the condensation of codes dealing with RS delivery trays and RS delivery tables. The third rater indicated that, although the tray and the table were different pieces of equipment that in some cases required different handling and preparation for usage, the spirit of the passage was regarding what or how something was to be delivered, and not that the delivery specifically utilized one piece of equipment over the other. The code for trays and tables, previously separated, now became one code (TRAYB) indicating the usage of either piece of equipment for the completion of the task.

Finally, many of the passages were replicated within each organization's documents. Specifically, a majority of the line items were expressed both in terms of the policy and whether it was established and clearly communicated, and in terms of action and whether the task was actually being completed as specified. Initially, both of the first two coders felt it was important to separate the managerial function of setting a standard or policy and its communication to the employees from the actual execution of the task by the staff. After some consideration, it was decided that these too should be collapsed, and by rewriting most passages with the starting phrase, "Are there policies in place that (task)," the nature of both functions were still considered. This supports the development of the audit as a self-assessment tool to be used by management and line staff alike, regardless of specific position or shift.

With these three types of approaches to data reduction completed, the initial two coders worked out any remaining or unique differences in the coding scheme. Grouping together similar passages that described different aspects of the same tasks developed the final condensation. For example, the passages may have separately indicated tasks such as cleaning silverware, cleaning plates, cleaning glasses, and cleaning the delivery tray or table itself, all in preparation for a delivery. It was determined that all tasks related in detail in the passage, but part of a larger overall chore, would be condensed, and that the specific sub-components of the task would be included as examples in italics in the initial draft of the generic audit. This also supports the study's goal of identifying best principles across different hotel organizations by allowing both the general and the detailed information to be visible, and ultimately, attuned to whichever facility uses the final audit version.

Although frequencies were generated for the study's coding by Atlas/ti, the coders did not consider them in the coding process. Since the goal of the study was to develop a comprehensive list from which to develop a generic audit that covers all types and styles of service, it was decided that all items that were not collapsed but had a frequency of only one, would remain in the first draft of the audit.

From this condensed draft of the initial coding an initial draft of the generic audit was developed that would serve as the instrument for review by the panel of experts.

The initial output from Atlas/ti was filled with connotations and codes that convoluted the interpretation by laymen unfamiliar with the program. So, a version of the audit was developed in Microsoft word that would clean up the data and present it in a digestible structure for the panel of experts in the Initial Audit Draft – Panel Version (Appendix B). The resulting initial audit draft fully condensed as previously described generated a 14-page document with 179 line items. These items were placed into a table. The full body of the text passage was rewritten as full and complete sentences with specifics as needed in the cell of the table. Smaller details were placed in the text body in italics. Headers were added to each super code category and each tier code category, and frequencies were placed by each line item and header for each category, reflecting the total number of codes for all items in that heading. This was done so that the panel of experts in the next phase of research could easily see the frequencies in order to help them determine the potential universality of each specific line item. Table columns were added for item number. Items on the audit were numbered using a mixed system of code and number. Since the data would be used in quantitative analysis, and due to the fact that it was far easier to refer to items with an identification number as opposed to the full passage of text, a straight numbering system was avoided, and the abbreviation of codes by section and type was adopted. The draft also added three columns after the text passage to be used as a self-assessment tool. These columns provided an area to acknowledge the existence of the policy, the completion of the policy, or the lack of applicability of that line item to the facility that was utilizing the tool. Finally, Appendix B was formatted to be easy to read and was paginated to avoid any line items across a page break.

4.2 Panel of Experts

A panel of experts was gathered for the next stage of the study. A convenience sample approach was used to solicit participants. The researcher contacted professional colleagues that he knew personally and explained the purpose and benefits of the study, as well as describing the time commitment and workload involved in participating. From willing and available participants, seven panel members and two alternates were selected. All panel members and alternates participated in the panel meetings, but the alternates provided their feedback and commentary in writing only, and did not participate in the oral conversation with the rest of the panel.

Panel members were purposively selected to represent a cross-section of management and staff that work or have worked in or with RS functions in first-class and luxury hotels. Participants were located throughout the U.S., and held the following positions in the hospitality industry; president of a hotel management company, director of human resources for a hotel management company, general manager of an independent resort, food and beverage director of a franchise hotel, executive chef for an independent hotel, room service chef for a franchise hotel, room service manager for a megaresort, and a room service waiter for a franchise hotel. Additionally, one associate professor of hospitality management who specializes in food and beverage participated. Collectively, the panel had more than 240 years of professional experience in the hospitality industry.

Arrangements were made with the panel to meet collectively via a conference call. Prior to the establishment of that date, all participants were sent a copy of the panel version of the generic audit, previously referred to as the initial audit draft. The researcher contacted all participants directly and personally in order to explain the duties of the panel. It was requested of each participant that they thoroughly review the audit file prior to the scheduled panel meeting. They were asked to specifically review the document for accuracy, clarity, redundancy, and completeness of all 179 line items in the audit. The researcher was interested in the feedback from the panel to assist in further simplifying the document while not losing the accuracy and level of detail in the line items. Additionally, participants were also asked to identify any salient practices that were not included in this draft of the audit. Figure 1 details the instructions sent to the panel.

Figure 1. Panel Directions.

Dear (Panel Member):

Thank you again for your participation in this room service research project. Your insight and feedback are extremely valuable to the outcome of this project. As we have discussed, there are three stages to the third project with which the panel is involved. These stages are: review, discussion, and ranking. Attached is the first draft of the room service generic audit. Please review the audit for room service operations in a first-class and luxury hotel using the following criteria:

- **Completeness:** Is this audit document representative of the tasks and functions necessary to conduct a room service operation appropriately and efficiently? Note where there are functions, duties, or other relevant criteria from your experiences in hospitality operations that are not fully related in this document.

- **Accuracy:** Are there functions, duties, or other operating standards for a room service department that are not included in this audit file? Please note items or processes that you believe are missing and are important to the operation.

- **Clarity:** Are the textual descriptions of the functions included in the audit understandable? Are they in an easy to read and format that different personnel at different levels in the room service operation will be able to follow and complete? Please make notes where the language could be revised to be clearer to the reader.

- **Redundancy:** Are there any items in the audit that are redundant? Are certain duties repeated that could be collapsed or rewritten into one larger categorical item? Please make notes regarding any such duplication.

Please feel free to contact me should you have any questions regarding this phase of the research. In a short while, you will be contacted to set up a date and time to meet in an online forum to discuss the notes or comments that developed from your review of the document. I thank you again for your cooperation and time.

A date was set that was convenient for all participants. Participants were called and added to a conference call so that all members could speak freely in real time. At the beginning of the meeting, all participants identified themselves to the panel, and the instructions were reviewed again. At this time, in addition to the accuracy, completeness, and clarity of the line items, participants were asked to provide additional feedback on the structure and format of the document itself, specifically in order to determine the optimal structure of the document to facilitate ease of use by RS staff during implementation. Alternates prepared their responses and commentary in writing, while the remaining participants were instructed to save such commentary for the panel meeting.

The panel conference call lasted more than three hours. The audit was reviewed aloud, line by line, and feedback was solicited after each item. The expertise and knowledge of the panel was of great benefit to the study. All participants were enthusiastic and contributed to the discussion. The collective knowledge of these professionals fulfilled their purpose by ensuring the line items were clear and concise, repetition of items were minimized, verbiage and syntax were appropriate and comprehensible to line staff, and that the format of the document was simple and effective for use during a work shift.

The panel felt that many of the items in the audit draft could be further condensed. Using the same approach as the initial coders, the panel focused on collapsing and condensing items they perceived as similar in order to further reduce the number of line items. They also revised the wording of several line items to be more concise and clear. After a thorough discussion of deletions, condensations, and revisions, the experts offered up multiple suggestions for additional line items they felt were missing from the initial draft. Some time was spent on the proper wording of these new passages to reflect the style of the items already in place.

It was suggested that the table itself be revised in several major ways. First, the panel felt the audit should be re-ordered. The initial version related line items by headings and sub-headings in alphabetical order. The panel suggested that it be reordered to reflect the actual chronology of the tasks over the course of a shift. They also suggested the shading of cells for greater figure-ground separation and ease of use. Page numbers were added. Adding a blank area after each line item for specific comments was suggested. A complete list of the revisions, additions, and format alterations is included in the Panel Revisions to the Audit Draft (Appendix C).

The panel greatly reduced the size of the document. The revised, streamlined document was reduced to 9 pages, with 85 line items, greater than a 50% reduction from the prior version, which contained 14 pages and 179 line items. The audit draft was revised to include all suggested revisions and additions. In preparation for the next phase of panel research, the format of the document was altered to replace the cells previously allocated for responding to the status of each line item with a 5-point Likert rating scale as shown in the Panel Rating Instrument (Appendix D).

The next phase of the panel research involved returning the revised audit draft, now referred to as the panel-rating version, to the panel for a final review and a rating of the importance of each line item. This was done to accomplish another goal of the study, which was to develop a list of best principles and practices in RS. The panel was again contacted personally and individually to discuss the directions for ranking the items. 70 Figure 2 shows the directions.

Figure 2 shows the directions that were also attached to the audit files that were sent to the panel. The 5-point Likert scale rated each item on its importance from 1 to 5, with 1 being of very low importance and 5 being of very high importance.

Figure 2. Panel Rating Directions.

Dear (Panel Member):

We continue to be grateful for your contribution to this research effort. This is the last step of your participation. I have enclosed a version of the generic Room Service Audit you helped to create and design.

Please rate each of the line items on the audit from 1 to 5. This can be done by printing out the 9-page document and simply circling the rating number provided for each line item in the cell following the textual description of the item, or you can type in the number selected in the final cell of each line digitally and return it via email to me directly. You will be rating these individual line items in accordance to their importance to the overall room service operation. The rating scale is as follows:

5 = the line item is of very high importance

4 = the line item is of high importance

3 = the line item is of moderate importance

2 = the line item is of low importance

1 = the line item is of very low importance

The next step of this research will be to survey the general population to look for differences in the perceived importance of the items you helped create and rank. The sooner we are able to collect your original rankings, the sooner we can move to this next and crucial step, so please try to complete this rating process at your earliest convenience. As always, please feel free to contact me via phone or email should you have any questions, comments, or concerns.

All panel members completed the ratings and returned them to the researcher within two weeks. The nine panel participants' ratings for each item were averaged and ranked in descending order of importance. All line items rated with an average above 4.0 (of high importance or greater) were selected to be identified as the best principles for RS. Since the research process thus far had identified all of items as important to RS operations, it was vital to determine which of these were seen as most important to the panel of experts. It was reasoned that any line item above moderate importance could be considered critical to the organization's successful execution of RS operations. There were 53 items of the 85 (62%) that were selected as the best practices in RS for the quantitative portion of the study. Table 2 shows the mean rating for each of the items in descending order.

4.3 Field Testing

While the panel was ranking the most important line items, a field test of the 85 line-item audit was performed. This field testing was conducted to lend a greater sense of face validity to the audit instrument. If line level staff could use the audit with ease and understanding, then it measures what it is supposed to, and has acceptable levels of face validity. The audit was revised, and the columns previous marked as Policy, Practice, and N/A were replaced with columns marked as Yes, No, and N/A in order to reduce confusion on whether the audit was being conducted to establish standards as opposed to verify that existing standards were being executed. The audit was also revised to add the previously suggested section for comments, and paginated to divide the functions so that separate sections could be handed to the employees whose duties and areas of responsibility matched with the audit sections. Figure 3 shows the complete field test version of the audit, including all revisions.

Table 2. Panel of Experts – Mean Ratings

Item	Mean Rating	Item	Mean Rating	Item	Mean Rating	Item	Mean Rating
AHR2	4.86	OTEGI2	4.29	POATT1	4.00	OPB1	3.57
AHR1	4.71	PPAI2	4.29	PPAS1	4.00	OPF1	3.57
APR1	4.57	RVIP3	4.29	RVIP2	4.00	POAL1	3.57
AHR3	4.57	AM2	4.14	ACG1	3.86	AI1	3.43
OPOR1	4.57	ODGI1	4.14	AM4	3.86	ODF1	3.43
APR2	4.43	ODGI3	4.14	AOH2	3.86	PPAI1	3.29
ACG3	4.43	ODVIP1	4.14	ODB1	3.86	RP2	3.29
AS3	4.43	OTEGI3	4.14	ODB2	3.86	RTO1	3.29
AS5	4.43	OTEGI4	4.14	ODGI6	3.86	AS6	3.14
ODA1	4.43	OTEGI7	4.14	POAS1	3.86	ODGI7	3.14
ODGI4	4.43	OTUS1	4.14	PPAC1	3.86		
OPF3	4.43	PPAC1a	4.14	RVIP1	3.86		
OTEGI1	4.43	PPAVIP1	4.14	ACCR1	3.71		
OTEGI6	4.43	RP1	4.14	AOH1	3.71		
OTPA1	4.43	ACG2	4.00	AS7	3.71		
APR3	4.29	AI2	4.00	ODGC2	3.71		
AM1	4.29	AS1	4.00	POAA1	3.71		
AM3	4.29	ODGI5	4.00	POAB1	3.71		
ODGD1	4.29	ODTS1	4.00	POATT2	3.71		
ODGI2	4.29	OPOR2	4.00	PPAC2	3.71		
OPB2	4.29	OPVIP1	4.00	RFD1	3.71		
OPF2	4.29	OTEGI5	4.00	ROP1	3.71		
OPTP1	4.29	OTPA2	4.00	AS2	3.57		
OPTP2	4.29	POAC1	4.00	AS4	3.57		
OTDE1	4.29	POATO1	4.00	ODGC1	3.57		

Figure 4 shows the directions that the researcher drew up for the field test. The researcher made arrangements with a large casino megaresort on the Las Vegas strip to conduct the field testing through RS employees over four separate visits to the property. The first meeting was with the hotel firm's corporate executive in charge of food and beverage. There, the audit was reviewed and the need for field testing was explained. fter agreeing to certain terms and conditions, including the anonymity of the facility, its employees, and any specifically unique practices in place in the RS department, contacts for the line level RS management and staff were provided.

Room Service Operating Audit – Field Test Version				
Number/Function		*Yes*	*No*	*N/A*
A	**Administrative**			
S	***Standards***			
AS-1	Are there standards established and in place for beverages? *Including coffee, hot tea variety, orange juice, liquor brands, beverage temperature, wine brands, coffee time on heat, accoutrements, iced tea, grapefruit juice, non-alcoholic beverages, beverage napkins, beverage promotions, water, preferred vendor, preferred brand.*			
Comments:				
AS-2	Are there standards established and in place for tableware? *Including candles, flatware, ice buckets, linen, plate covers, pots, ashtrays, salt and pepper shakers, bud vases.*			
Comments:				
AS-3	Are there standards established and in place for food products? *Including Fruit, Vegetables, Breads, yogurt, cereal, eggs.*			
Comments:				
AS-4	Are there standards established and in place for condiments? *Including jam, salad dressing, ketchup, mustard, salt, sugar, syrup, portion size.*			
Comments:				
AS-5	Are there standards established and in place for employee uniforms? *Including name tag, brand compliance, appropriateness to region/market.*			
Comments:				
AS-6	Are there standards established and in place for dairy products? *Including butter, butter temperature, cream, milk.*			
Comments:				
AS-7	Are there standards established and in place for delivery trays and tables? *Including tray liners, tray/table set-up, tray material.*			
Comments:				

Room Service Operating Audit – Field Test Version				
Number/Function		*Yes*	*No*	*N/A*
HR	***Human Resources***			
AHR-1	Are policies in place for employee/guest interaction? *Including employee attitude, using guest name, greeting guests, use of appropriate language, accuracy of information, asking customers to wait, providing guest directions, selling of facility services, volume of voice, unavailability of facility services.*			
Comments:				
AHR-2	Are policies in place for employee training? *Including beverage, hotel knowledge, menu, wine service, RS dispatcher, guest privacy, employee/guest gender sensitivity.*			
Comments:				
AHR-3	Are policies in place for the appearance of employees? *Including jewelry, hygiene, hair, facial hair, fingernails.*			
Comments:				
CG	***Condition of Goods***			
ACG-1	Are policies in place for the condition of tableware? *Including glassware, china, flatware, linen, pots, sugar bowls, candles.*			
Comments:				
ACG-2	Are policies in place for the condition of RS delivery trays and tables?			
Comments:				
ACG-3	Are policies in place for the condition of employee uniforms?			
Comments:				
M	***Menu***			
AM-1	Are there policies for menu content and design? *Including menu variety, culinary trends, operating hours, menu accuracy, menu item descriptions, language(s) of menu, dietary and nutritional concerns, outlets and areas of hotel covered by room service.*			
Comments:				
AM-2	Are there policies in place for the availability of menu items? *Including children's items, no out of stock items, quick-service items, full beverage list, seasonally available products, market specific products, preparation of non-menu items.*			
Comments:				

Room Service Operating Audit – Field Test Version				
Number/Function		*Yes*	*No*	*N/A*
AM-3	Are there policies in place for menu pricing? *Including children's prices, prices kept current, prices appropriate to market.*			
Comments:				
AM-4	Are there policies in place for menu restrictions by location in the hotel? *Including areas with no delivery permitted, areas where alcoholic beverages are restricted, areas where glassware is prohibited.*			
Comments:				
PR	***Problem Resolution***			
APR-1	Are there policies in place for employee/guest interaction during PR? *Including apologizing to guest, active listening, employee attitude, estimated timing of resolution, thanking the customer, which staff member handles these calls.*			
Comments:				
APR-2	Are there policies in place for the empowerment of staff to resolve guest problems?			
Comments:				
APR-3	Are there policies in place for calling back a guest during PR? *Including estimated timing of resolution, following up to ensure satisfaction, which staff member handles these calls.*			
Comments:				
I	***Inventory***			
AI-1	Are there policies in place for the par levels of supply inventory? *Including beverage, china, food, linen, glassware, flatware.*			
Comments:				
AI-2	Are there policies in place for the timing of physical inventories? *Including beverage, food, supplies, which staff members conduct inventory.*			
Comments:				
CCR	***Cost Controls-Receiving***			
ACCR-1	Are policies in place for receiving and invoicing of products?			
Comments:				

Room Service Operating Audit – Field Test Version				
Number/Function		*Yes*	*No*	*N/A*
OH	*Operating Hours*			
AOH-1	Are there policies in place for minimum operating hours?			
Comments:				
AOH-2	Are there policies in place for the posting of operating hours in the hotel? *Including estimated timing of delivery.*			
Comments:				
PPA	**Pre-Prep Activities**			
S	*Stocking*			
PPAS-1	Are there policies in place for the stocking of supplies? *Including bread, sugar bowls, tea, butter, jam, lemon, coffee, dairy products, flowers, ice, iced tea, juice, newspapers, folded napkins, salt and pepper, syrup, linen, china, flatware, glassware, trashcans, trays/tables with condiments.*			
Comments:				
C	*Cleaning*			
PPAC-1	Are there policies in place for the cleaning of tableware? *Including butter containers, creamers, china, flatware, glassware, jam containers, syrup containers, salt and pepper shakers.*			
Comments:				
PPAC-2	Are there policies in place for the cleaning of FF&E? *Including coffee, tea, bread, juice, trays/tables.*			
Comments:				
I	*Inspection*			
PPAI-1	Are there policies in place for inspecting the trays/tables?			
Comments:				
PPAI-2	Are there policies in place for checking expiration dates? *Including dairy products, flowers, syrup.*			
Comments:				
C	*Communication*			
PPAC-1	Are there methods in place for communicating pertinent information? *Including message logs, memos, management conversation, meetings, specials, out-of-stock items.*			
Comments:				

Room Service Operating Audit – Field Test Version				
Number/Function		*Yes*	*No*	*N/A*
VIP	*VIP Orders*			
PPAV IP-1	Are there policies in place for verifying amenity order information?			
Comments:				
OT	**Order Taking**			
EGI	*Employee/Guest Interaction*			
OTEG I-1	Are there policies in place for the verification of the guestroom order? *Including guestroom number, number of guests, guest order, who takes calls, how calls are routed.*			
Comments:				
OTEG I-2	Are there policies in place for the usage of a guests' name?			
Comments:				
OTEG I-3	Are there policies in place for placing a guest on hold? *Including asking permission, thanking the guest for holding, estimating the length of time on-hold, offering a call-back as an alternative to holding.*			
Comments:				
OTEG I-4	Are there policies in place for the quality of the dispatcher's speech?			
Comments:				
OTEG I-5	Are there policies in place for offering assistance during the ordering process?			
Comments:				
OTEG I-6	Are there policies in place for employee attitude and behavior?			
Comments:				
OTEG I-7	Are there policies in place for thanking the guest for their order?			
Comments:				
PA	*Phone Answering*			
OTPA -1	Are there policies in place for the timing of answering a call?			
Comments:				

Room Service Operating Audit – Field Test Version				
Number/Function		*Yes*	*No*	*N/A*
OTPA -2	Are there policies in place for the identification of the employee name?			
Comments:				
US	***Up-selling***			
OTUS -1	Are there policies in place for the up-selling of RS products? *Including beverage, food.*			
Comments:				
DE	***Delivery Estimation***			
OTDE -1	Are there policies in place for the estimation of delivery time? *Including guarantees, prioritization of orders by type.*			
Comments:				
R	**Routing**			
VIP	***VIP Orders***			
RVIP-1	Are there policies in place for the processing of a VIP order? *Including logging of order, labeling of order, filling of order.*			
Comments:				
RVIP-2	Are there policies in place for the cancellation of an amenity order? *Including communication of cancellation to personnel, logging of cancelation, return of amenity to person who ordered it.*			
Comments:				
RVIP-3	Are there policies in place for the verification of a VIP order? *Including order accuracy, guest name, room number.*			
Comments:				
OP	***Order Processing***			
ROP-1	Are there policies in place for the assignation of an order to a server?			
Comments:				
P	***Posting***			
RP-1	Are there policies in place for the posting of orders to the POS? *Including beverage, flowers, bottled water, entire order.*			
Comments:				

Room Service Operating Audit – Field Test Version		Yes	No	N/A
Number/Function		*Yes*	*No*	*N/A*
RP-2	Are there policies for random verification of orders posted to the POS? *Including beverage, flowers, entire order.*			
Comments:				
TO	***Timed Orders***			
RTO-1	Are there policies for random verification of orders posted to the POS? *Including beverage, flowers, entire order.*			
Comments:				
FD	***Flower Delivery***			
RFD-1	Are there policies in place for the processing of flower orders? *Including authorization, logging of order, price, routing of order to server.*			
Comments:				
OP	**Order Preparation**			
TP	***Table/Tray Preparation***			
OPTP-1	Are there policies in place for the preparation of a delivery tray/table? *Including condiments, china, tray/table retrieval information, hot food, glassware, flatware, after-dinner mints with dinner service, beverages, flowers, tray liner.*			
Comments:				
OPTP-2	Are there policies in place for the overall appearance of the delivery tray/table?			
Comments:				
B	***Beverage***			
OPB-1	Are there policies in place for the preparation of beverage accoutrements? *Including tea, coffee, soda.*			
Comments:				
OPB-2	Are there policies in place for the measurement of alcoholic beverages?			
Comments:				
F	***Food***			
OPF-1	Are there policies in place for the garnishing of food?			
Comments:				

Room Service Operating Audit – Field Test Version				
Number/Function		*Yes*	*No*	*N/A*
OPF-2	Are there policies in place for the covering of hot food with plate covers?			
Comments:				
OPF-3	Are standardized recipes in place for the production of room service menu items? *Including access to written standard recipes, training of culinary personnel, menu review, menu engineering.*			
Comments:				
OR	***Order Review***			
OPOR -1	Are there policies in place for reviewing the order for accuracy prior to exiting the kitchen? *Including who conducts review.*			
Comments:				
OPOR -2	Are there policies in place for the cleaning of delivery supplies? *Including plate covers, china, glassware, flatware.*			
Comments:				
VIP	***VIP Orders***			
OPVI P-1	Are there policies for the preparation and prioritization of a VIP order? *Including china, flatware, linen, tableware, order type, priority of orders.*			
Comments:				
OD	**Order Delivery**			
GI	***Guest Interaction***			
ODGI -1	Are there policies in place for asking about guest dining preferences? *Including where the guest would like to sit, where the guest would like the tray or table set up, what condiments the guest would prefer.*			
Comments:				
ODGI -2	Are there policies in place for thanking the guest?			
Comments:				
ODGI -3	Are there policies in place for the usage of a guest's name?			
Comments:				

Room Service Operating Audit – Field Test Version				
Number/Function		*Yes*	*No*	*N/A*
ODGI -4	Are there policies in place for a review and description of the delivery for accuracy? *Including verbal repeat of the order to the guest.*			
Comments:				
ODGI -5	Are there policies in place for informing the guest of tray retrieval policies?			
Comments:				
ODGI -6	Are there policies in place for the mentioning or not mentioning gratuity inclusion on a guest check?			
Comments:				
ODGI -7	Are there policies in place for advising the guest of safe and proper use of room service equipment?			
Comments:				
GC	***Guest Checks***			
ODGC-1	Are there policies in place for the presentation of the guest check? *Including use of folder, inclusion of pen, when to present, check branding, comment card.*			
Comments:				
ODGC-2	Are there policies in place for the presence of a service charge? *Including location and visibility of service charge on guest check.*			
Comments:				
TS	***Table Setting***			
ODTS -1	Are there policies in place for setting the guest table? *Including linen, chairs, china, condiments, flatware, glassware, flowers.*			
Comments:				
B	***Beverage***			
ODB-1	Are there policies in place for wine service?			
Comments:				
ODB-2	Are there policies in place for uncapping and pouring beverages? *Including who serves, alcoholic beverages.*			
Comments:				

Room Service Operating Audit – Field Test Version				
Number/Function		*Yes*	*No*	*N/A*
F	***Food***			
ODF-1	Are there policies in place for the uncovering of food?			
Comments:				
GD	***Guestroom Door***			
ODGD-1	Are there policies in place for guestroom door usage? *Including closing, opening, requesting entry, do not disturb signage, securing door in open position for employee safety.*			
Comments:				
A	***Announcement***			
ODA-1	Are there policies in place for the announcement of an RS delivery? *Including identification of employee, use of guest name.*			
Comments:				
VIP	***VIP Orders***			
ODVIP-1	Are there policies in place for the preparation and delivery of VIP orders? *Including timing of deliveries, set up of orders in guestroom, inclusion of amenity card, type of amenity.*			
Comments:				
POA	**Post-Order Activities**			
TT	***Trays/Tables***			
POATT-1	Are there policies in place for the retrieval of guest room trays/tables? *Including cross-departmental collaboration, transfer of retrieval duties to other departments, message logs, memos, management conversation, meetings.*			
Comments:				
POATT-2	Are there policies in place for the return of trays/tables to their assigned location? *Including cross-departmental collaboration, transfer of retrieval duties to other departments.*			
Comments:				

Room Service Operating Audit – Field Test Version				
Number/Function		*Yes*	*No*	*N/A*
TO	***Timed Orders***			
POAT 0-1	Are there policies in place for the offering, availability, and collection of timed orders?			
Comments:				
B	***Breakdown***			
POAB -1	Are there policies in place for breakdown and recovery of RS food stock? *Including butter, cream, jam, juice, syrup.*			
Comments:				
C	***Cleaning***			
POAC -1	Are there policies in place for the cleaning of stations and equipment? *Including bread workstation, coffee workstation, tea workstation, delivery table hot boxes.*			
Comments:				
L	***Linen***			
POAL -1	Are there policies in place for linen preparation? *Including linen sorting, linen inspection.*			
Comments:				
A	***Amenities***			
POAA -1	Are there policies in place for the filing of delivered amenities? *Including type of amenity.*			
Comments:				
S	***Stocking***			
POAS -1	Are there policies in place for the stocking of needed supplies? *Including condiments, salt and pepper, sugar.*			
Comments:				

Figure 3. Final Audit Draft – Field Testing.

Figure 4. Room Service Field Test Directions.

Thank you for taking the time to review the following operating audit for room service operations.

Please review the section(s) given to you. The researchers are interested in your feedback on this audit. Specifically, provide your comments and insight on the accuracy, clarity, completeness or redundancy of each line item. In general, is the audit easy to use? Does it accurately describe the various activities required of a room service operation? Are the duties of the service(s) you are involved with complete? Is there anything missing? We will address these issues either while you are exploring your section or when you are done.

Finally, does the audit itself make sense? Is there adequate space to add commentary on specific line items? Does the Yes, No, and N/A columns make sense? Please feel free to question the researcher at any time during your review of this document. Thank you again for your time, effort, and consideration.

The second visit to the property provided a chance for the researcher to meet the RS manager and her assistants and shift supervisor. There was great interest in the study from the RS management. They assisted the researcher in identifying which areas and duties pertained to which staff positions, and helped further divide the audit into appropriate sections. A list of employees and their work schedule for a two-week period was provided. The researcher reviewed the schedule and staffing levels and selected an evening and a morning shift to observe and facilitate the audit process at the property. A set of directions for Figure 3 was generated and provided to the RS management. RS management held a meeting on behalf of the researcher with the entire RS staff. At this meeting, appropriate sections of the audit, along with the directions were handed to each employee. The employees were urged to review the document and be prepared to utilize the audit on their next schedule shift. Employees working on the two shifts scheduled and approved by the RS management were directed to wait to complete the audit until the researcher was with them.

The next two visits to the property entailed observation of the RS staff conducting their duties and checking their audits against their tasks. As an added surprise, between the meeting and the schedule shifts, RS management went through the generic file and added in the property specific standards for each line item. Management noted to the researcher that the standards were supposed to be established, well communicated, and in place, but was concerned that the generic state of the document, absent of the actual quantitative and property specific standards, might confuse some of the staff. During each visit to the actual operation, RS staff seemed at ease with the audit process. Some asked occasional questions about why this was conducted, and where the information came from, but not one employee requested further clarification of the line items.

The researcher did not conduct the audit himself. Rather the staff conducted the audit portions assigned to them and the researcher observed, making notes, or inquiring about specificities as they executed their tasks. The audit was designed to be a selfassessment tool, and the field test's chief goal was to see it in use and ensure it made sense to those who used it. After each 8 hour shift, the researcher collected the completed audits. The RS management provided the remainder of the completed audits from staff that were not scheduled to work on the researcher's two shifts within a week after the second shift. All audits were completed in full. Though there were many notes provided in the comment section for each line item, none of the feedback referred to a lack of understanding of the line item itself. All comments were directed towards why a particular standard may have failed, or why some standards were not precisely accurate. In general, the audit field testing went smoothly and without incident. The staff and management were pleasant, professional, and cooperative. A few weeks after the field testing was completed, the researcher received a letter from a newly appointed RS manager that thanked the researcher for the visits to the property, and related that the facility had adopted the self-assessment tool for regular usage by staff. She added that the staff enjoyed using the tool and saw it as a source of empowerment in addition to its efficacy as a tool for ensuring important duties are not overlooked or completed poorly.

4.4 MANOVA

An analysis of variance was conducted on the 53 line items within the final draft of the audit that were identified as most important to the success of a RS operation by the panel of experts. This study utilized an online service, Qualtrics, to gather the data.It is a market research firm that offers an online platform for gathering real-time market intelligence and insights for academic and industry analysis. Qualtrics' services for this research included the posting and maintaining of the survey instrument, the identification and recruitment of respondents from their national industry databases, and the storage of the data from those respondents.

The survey instrument was posted online utilizing the exact same verbiage and format as the instrument utilized by the panel of experts. Only the 53 items rated as most important by the panel were used. The instrument also asked additional questions. The first added question was a qualifying question. It asked whether the respondent had hospitality food and beverage experience. The researcher felt that since there was such a high level of detail in the text of the line items, that it would require individuals with a level of familiarity with the terminology and the general workings of food and beverage operations within a hotel. If respondents indicated that they did not have a hospitality food and beverage experience, they were thanked for their time and were not allowed to continue the survey. All other respondents who indicated they had such experience were able to continue with the online survey.

Four additional questions were added to the survey to assist with the MANOVA portion of this study. These four questions were demographic in nature, and corresponded with the four hypotheses this study investigated. These questions were concerned with the types of facilities with which the respondents had professional work experience. These questions inquired about the size of the hotel (200 guest rooms or less vs. 201 guest rooms or more), the type of hotel (resort vs. non-resort), the type of room service operation (free-standing department vs. a function within an existing outlet), and the franchise status of the hotel (franchised vs. independent). After completing these questions, the respondents then continued on with rating the 53 audit line-items based on perceived importance of the item, on a scale from 1 to 5, with 1 being of very low importance, and 5 being of very high importance. The questions utilized the same scale and the same line item text as the instrument developed with and utilized by the panel of experts.

Qualtrics contacted over 1000 pre-qualified individuals with appropriate professional backgrounds from their industry databases to solicit participation in this study. Of those that did participate, 302 qualified to complete the study, and those 302 completed the entire study. Results from this study were analyzed using SPSS (version 12.0), a statistical software program commonly used for academic statistical analysis research such as this project.

The study proposed four hypotheses:

H1: There is no difference in the perceived importance of key room service practices between employees of a resort hotel and employees of a non-resort hotel.

H2: There is no difference in the perceived importance of key room service practices between employees of a large hotel and employees of a small hotel.

H3: There is no difference in the perceived importance of key room service practices between employees of a hotel with a freestanding room service department and employees of a hotel without a freestanding room service department.

H4: There is no difference in the perceived importance of key room service practices between employees of a franchise hotel and the employees of an independent hotel.

Testing for each of these hypotheses is based on a 2-factor MANOVA. In other words, each hypothesis compared to groups: one that possessed the characteristic; and one that did not. In this study, these characteristics were based on whether or not the respondent possessed work experience at a specific type of hotel room service operation. In Yes/No queries such as these, where there are only two factors, post hoc tests are not necessary (Muhr, 2005). In two-factor MANOVA studies, the most important assumption is that of homogeneity of variance, also referred to as equality of variance. SPSS utilizes Box's Test of Equality of Covariance Matrices to investigate the equality of variance between the two respondent groups. For each hypothesis, Box's test indicated that the assumption of equality of variance between the two groups was violated. Box's test is very sensitive and is often violated in such research studies. According to the SPSS manual (Norusis, 2004), the sensitivity of Box's test can be countered by a large enough sample size. In this case, a 302 respondent sample was large enough to disregard the violation of the assumption of equality of variance.

H1

The first hypothesis (H1) investigated the differences in the perceptions of importance of the 54 RS operating procedures between respondents with professional experience at a resort hotel and those with professional experiences at a non-resort hotel. So, a null and an alternate hypothesis were developed:

$H1^{o}$: There is no difference in the perceived importance of key room service practices between employees of a resort hotel and employees of a non-resort hotel.

$H1^{a}$: There is a difference in the perceived importance of key room service practices between employees of a resort hotel and employees of a non-resort hotel.

For the first hypothesis, there were 302 respondents (n = 302) that qualified to participate in the study and that fully completed the online survey instrument. Table 3 shows that of the 302 respondents, 133 (44.1%) had the majority of their work experience in resort hotels while 169 (55.9%) had the majority of their work experience at a nonresort hotel.

Table 3. Descriptive Statistics - H1

Between-Subjects Factors

		Value Label	N
At what type of hotel has the majority of your work experience been?	1	Resort Hotel	133
	2	Non-Resort Hotel	169

As shown in Table 4, Box's test was violated. A significance level of less than .05 would normally indicate a violation of the assumption of homogeneity. However, in this study, the sample size (n = 302) was large enough to disregard the violation and move to determine whether the null hypothesis was to be rejected or fail to be rejected.

Table 4. Box's Test – H1

Box's Test of Equality of Covariance Matrices	
Box's M	2305.804
F	1.705
df1	1128
df2	242303.525
Sig.	.000

The multivariate analysis was conducted utilizing SPSS software. Wilks' Lambda was computed at an alpha level of .05. Wilks' Lambda for this output, located in Table 5 in the section labeled Q2.2 was .052, slightly above the alpha level of .05.

Table 5. Multivariate Tests – H1

Effect		Value	F	Hypothesis df	Error df	Sig.	Partial Eta Squared	Noncent. Parameter	Observed Power
Q2.2	Pillai's Trace	.207	1.407	47.000	254.000	.052	.207	66.111	.995
	Wilks' Lambda	.793	1.407	47.000	254.000	.052	.207	66.111	.995
	Hotelling's Trace	.260	1.407	47.000	254.000	.052	.207	66.111	.995
	Roy's Largest Root	.260	1.407	47.000	254.000	.052	.207	66.111	.995

A .052 Wilks' Lambda level indicates that there were no significant differences in the perceptions of importance of RS operating policies between respondents with professional experience in a resort hotel and those with professional experience in a nonresort hotel. Therefore, for H1, we fail to reject the null hypothesis that there is no difference between these two groups.

H2

The second hypothesis (H2) investigated the differences in the perceptions of importance of the 54 RS operating procedures between respondents with professional experience at a large hotel and those with professional experiences at a smaller hotel. 200 guest rooms was the cut off for the determination of large versus small. A small hotel had 200 or fewer guest rooms, whereas a large hotel had 201 or more guest rooms. So a null and an alternate hypothesis were developed:

H2^{o}: There is no difference in the perceived importance of key room service practices between employees of a large hotel and employees of a small hotel.

H2^{a}: There is a difference in the perceived importance of key room service practices between employees of a large hotel and employees of a small hotel.

For the second hypotheses, there were 302 respondents (n = 302) that qualified to participate in the study and that fully completed the online survey instrument. Of the 302, 155 (51.3%) had the majority of their work experience in smaller hotels, while 147 (48.7%) had the majority of their work experience at larger hotels (Table 6).

Table 6. Descriptive Statistics – H2

Between-Subjects Factors

		Value Label	N
At what size property has the majority of your work experience been?	1	1-200 Guestrooms	155
	2	201 or more Guestrooms	147

As stated previously, Box's test was violated (Table 7). A significance level of .000 (less than significance level of .05) would normally indicate a violation of the assumption of homogeneity. However, in this study, the sample size (n = 302) was large enough to disregard the violation and move to determine whether the null hypothesis was to be rejected or fail to be rejected.

The multivariate analysis was conducted utilizing SPSS software (Table 8). Wilks' Lambda was computed at an alpha level of .05.

Wilks' Lambda for this output, located in the table in the section labeled Q2.3 was .468, considerably above the alpha level of .05.

Table 7. Box's Test – H2

Box's Test of Equality of Covariance Matrices	
Box's M	2250.801
F	1.671
df1	1128
df2	269379.322
Sig.	.000

Table 8. Multivariate Tests – H2

Effect		Value	F	Hypothesis df	Error df	Sig.	Partial Eta Squared	Noncent. Parameter	Observed Power
Q2.3	Pillai's Trace	.157	1.007	47.000	254.000	.468	.157	47.317	.950
	Wilks' Lambda	.843	1.007	47.000	254.000	.468	.157	47.317	.950
	Hotelling's Trace	.186	1.007	47.000	254.000	.468	.157	47.317	.950
	Roy's Largest Root	.186	1.007	47.000	254.000	.468	.157	47.317	.950

A .468 Wilks' Lambda level indicates that there were no significant differences in the perceptions of importance of RS operating policies between respondents with professional experience in a larger hotel and those with professional experience in a smaller hotel. Therefore, for H2, we fail to reject the null hypothesis that there is no difference between these two groups.

H3

The third hypothesis (H3) investigated the differences in the perceptions of importance of the 54 RS operating procedures between respondents with professional experience at a hotel with a free-standing room service department and those with professional experience at a hotel with no free-standing room service department. A freestanding room service department has its own staff, and often its own production and staging facility separate from other departments in the food and beverage division of the hotel. A non free-standing department utilizes existing production facilities that were chiefly designed for another reason, such as a restaurant or banquet kitchen. So, a null and an alternate hypothesis were developed:

$H3^{o}$: There is no difference in the perceived importance of key room service practices between employees of a hotel with a freestanding room service department and employees of a hotel without a freestanding room service department.

$H3^{a:}$ There is a difference in the perceived importance of key room service practices between employees of a hotel with a freestanding room service department and employees of a hotel without a freestanding room service department.

For the third hypotheses, there were 302 respondents (n = 302) that qualified to participate in the study and that fully completed the online survey instrument. Of the 302, 187 (61.9%) had the majority of their work experience in resort hotels, while 115 (38.1%) had the majority of their work experience at a non-resort hotel (Table 9).

As stated previously, Box's test was violated (Table 10). A significance level of .000 (less than significance level of .05) would normally indicate a violation of the assumption of homogeneity. However, in this study, the sample size (n = 302) was large enough to disregard the violation and move to determine whether the null hypothesis was to be rejected or fail to be rejected.

Table 9. Descriptive Statistics – H3

Between-Subjects Factors			
		Value Label	N
At the hotel(s) where the majority of your work experience has been, was the room service function at...	1	Free-standing Room Service Department	187
	2	No Free-standing Room Service Department	115

Table 10. Box's Test – H3

Box's Test of Equality of Covariance Matrices	
Box's M	2211.635
F	1.613
df1	1128
df2	179516.508
Sig.	.000

The multivariate analysis was conducted utilizing SPSS software (Table 11). Wilks' Lambda was computed at an alpha level of .05. Wilks' Lambda for this output, located in the table in the section labeled Q2.5 was .109, well above the alpha level of .05.

A .109 Wilks' Lambda level indicates that there were no significant differences in the perceptions of importance of RS operating policies between respondents with professional experience in a hotel with a free-standing room service department and those with professional experience in a hotel with no free-standing room service department. Therefore, for H2, we fail to reject the null hypothesis that there is no difference between these two groups.

Table 11. Multivariate Tests – H3

Effect		Value	F	Hypothesis df	Error df	Sig.	Partial Eta Squared	Noncent. Parameter	Observed Power
Q2.5	Pillai's Trace	.193	1.294	47.000	254.000	.109	.193	60.809	.990
	Wilks' Lambda	.807	1.294	47.000	254.000	.109	.193	60.809	.990
	Hotelling's Trace	.239	1.294	47.000	254.000	.109	.193	60.809	.990

H4

The fourth hypothesis (H4) investigated the differences in the perceptions of importance of the 54 RS operating procedures between respondents with professional experience at a franchised hotel and those with professional experience at an independent hotel. A franchised hotel has an established brand and is supported by a corporate office that makes policy and standards decisions that help to reflect a specific brand image. An independent hotel has no established franchise standards from which it can draw or review to build its operating processes. So, a null and an alternate hypothesis were developed:

$H4^{o}$: There is no difference in the perceived importance of key room service practices between employees of a franchise hotel and the employees of an independent hotel.

$H4^{a}$: There is a difference in the perceived importance of key room service practices between employees of a franchise hotel and the employees of an independent hotel.

For the fourth hypotheses, there were 302 respondents (n = 302) that qualified to participate in the study and that fully completed the online survey instrument. Of the 302, 179 (59.3%) had the majority of their work experience in franchised hotels, while 123 (40.7%) had the majority of their work experience at independent hotels (Table 12).

Table 12. Descriptive Statistics – H4

Between-Subjects Factors

		Value Label	N
At what type of property has the majority of your work experience been?	1	Brand-Affiliated or Franchised Hotel	179
	2	Independent Hotel	123

As stated previously, Box's test was violated (Table 13). A significance level of .000 (less than significance level of .05) would normally indicate a violation of the assumption of homogeneity. However, in this study, the sample size (n = 302) was large enough to disregard the violation and move to determine whether the null hypothesis was to be rejected or fail to be rejected.

The multivariate analysis was conducted utilizing SPSS software (Table 14). Wilks' Lambda was computed at an alpha level of .05. Wilks' Lambda for this output, located in the table in the section labeled Q2.5 was .003, below the alpha level of .05.

A .003 Wilks' Lambda level indicates that there were significant differences in the perceptions of importance of RS operating policies between respondents with professional experience in a franchised hotel and those with professional experience in an independent hotel. As an additional note of support and because the Box's test for equality of variance is violated (even though the sample size is robust enough to disregard such a violation), Pilia's Trace can be investigated as well. Pilia's Trace was at .003, also below the alpha level of .05. Therefore, for H4, we reject the null hypothesis that there is no difference between these two groups.

Table 13. Box's Test – H4

Box's Test of Equality of Covariance Matrices	
Box's M	2302.340
F	1.691
df1	1128
df2	209052.855
Sig.	.000

Table 14. Multivariate Tests – H4

Effect		Value	F	Hypothesis df	Error df	Sig.	Partial Eta Squared	Noncent. Parameter	Observed Power
Q2.4	Pillai's Trace	.246	1.761	47.000	254.000	.003	.246	82.757	1.000
	Wilks' Lambda	.754	1.761	47.000	254.000	.003	.246	82.757	1.000
	Hotelling's Trace	.326	1.761	47.000	254.000	.003	.246	82.757	1.000
	Roy's Largest Root	.326	1.761	47.000	254.000	.003	.246	82.757	1.000

It was determined that there were significant differences in the perception of importance of room service operating policies between professionals with work experience in a franchised hotel and professionals with work experience in an independent hotel. SPSS output allowed for the identification of which of the individual line items in the instrument were found to be of significant difference between these two groups. The test of between-subject effects (Appendix E) identified the significance level of the differences between franchised and independent hotel employee perceptions for each line item individually. Any line item with significance below the established alpha level of .05 was considered to be significantly different in terms of perceived importance. An examination of the column labeled Sig. in the Test of Between-Subjects Effects reveals the significance level for each of the audit line items perceived difference between the two groups.

In order to more easily investigate which policies were seen to be of different perceived importance among franchised and independent hotels, the line items were revised from the full textual description of the policy to an abbreviated terminology. The descriptive statistics output (Appendix F) for this hypothesis allowed the identification of which group gave a higher rating to each of the items that were perceived significantly differently.

Since the online survey was reduced after the panel-rating phase, the original item numbering is no longer relevant to this output and would be very likely to confuse the reader due to the numerous versions of the audit that were created and revised during the research process. Items are now simply listed in the order they appeared in the survey, and the item was identified with a key word or phrase from that line item. This allowed for ease in cross referencing the data amongst its various formats. To that data was added the significance in difference and the specific mean rating from each group, derived from the output from the Test of Between Subjects Effects and the Descriptive Statistics. All of this data was converted to a new output table (Appendix G).

Of the 53 line items, 21 (39.6%) were found to have significant difference in the perception of their importance by individuals with work experience in a franchised hotel as compared to individuals with work experience in an independent property. The franchise hotel group rated all 21 of the significantly different items higher. The independent hotel group rated none of the significantly different items higher.

4.4 Overall Rating

One of the goals of this study was to identify the most vital and salient room service operating procedures. The panel of experts assisted in condensing the list to 53 items through their various meetings and feedback. These were the items that were rated by the online respondents. Using the same approach as the rating completed by the panel, a list of all items rated was constructed in descending order of mean ratings by the 302 online survey respondents (Appendix H) on a 1-5 scale.

This study utilized the same criteria as it did with the panel of experts to determine which of the policies for room service operation were seen as most vital. On the 5-point Likert rating scale, 4 signified of high importance and 5 signified of very high importance. It was arbitrarily determined that any item rated 4.0 or higher was among the most important room service operating procedures. This cut-off yielded 27 (50.1%) items, which could be considered among the most important RS procedures for a firstclass and luxury hotel.

The survey results section reviewed the processes behind the identification and development of the coding schema as well as the actual coding process. It provided a detailed overview of the contributions made to the audit instrument's structure, format, and content by the panel of experts. It examined the evolution of the generic audit instrument during and through every step of the multimethod data collection and analysis process, including its application and interpretation by actual hotel room service employees during the field-testing phase. Finally, Chapter 4 examined the data from the online survey using quantitative software to assist in verifying the significance of the data with respect to variations in perceived importance by hospitality food and beverage employees and managers. The data gathered from the qualitative portions of this study, combined with the output from the quantitative portions generated very rich results. A further interpretation of some of the more noteworthy results follows in the final discussion section.

PART V

5. DISCUSSION

5.1 Summary and Discussion

This study posed four research questions. Those research questions, presented in the first section of this study guided the development of the methodology, and kept the immense amount of data gathered during the study focused on that data that most directly assisted in answering the original research questions. Those research questions were:

- What are the most vital and salient characteristics of a successful, full-service, inroom dining program?
- Which operating processes can be identified as foundational practices necessary to ensure a quality experience for the guest?
- Is it possible to create an overview of these identified practices that would address all, if not most, types of full-service first-class/luxury hotels?
- What are the noteworthy, structural differences between these standards across company or asset types?

This section focused on each of these questions individually, and discussed the results from the five stages of research that generated the data analyzed in the prior section. Additionally, this section discussed the implications of those findings in terms of their significance, both practically and academically. Because one of the main goals of this study was to define what operating practices were currently in place in first-class and luxury hotel room service operations, this section also examined the applications of the generic room service audit that the study developed as a result of the input from various parties during the data gathering stages of the study. Any noteworthy limitations and any notes of specific interest that arose during the study were presented throughout the discussion of the results as they relate to the original four research questions and resultant hypotheses. Finally, this section will examine and recommend future research projects that would logically proceed from these findings.

5.2 Research Question 1

The first question asked which practices in room service in first-class and luxury hotels were the most vital. The entire structure of this multimethod study considered that query repeatedly. The study first reviewed the literature to gain a larger understanding of the general steps of a room service operation. The review also considered the contributions of hospitality textbooks and trade articles in establishing a primary and general sequence of events necessary to conduct room service operations. From this literature review, the initial coding schema arose, with the practical components and the sequence of events creating the major list of tier one family or super codes. These codes provided the framework and organized the perspective of the coding approach for the next stage of the study, the content analysis.

During the content analysis, over 1000 pages of corporate historical documents were coded and collated into categories arising first from the literature, and then from the content of the operating documents themselves. These documents were generated from several different structural types of hospitality organizations, including franchised properties, independent properties, and properties operated by hotel management firms that specialize in the operations of hotels for third parties. The first round of content analysis yielded over 660 coded passages contained in over 200 line items. It could be considered that all of these items are seen as important by some hospitality management entity that took the time to determine what operating aspect needed to be defined and/or implemented in the first place, thus creating its textual inclusion in the original historical documents.

A goal of qualitative content analysis is to represent a specific population as opposed to generalize findings from a small group to a larger population. Representing every specific unique operating behavior that a hotel utilizes, often to distinguish itself from its competition was not the goal of this study. Rather, it was to determine which functions in a room service operation that operators commonly or most often used in the first-class and luxury hotel segment. At a certain point, the bulk of the content and coding process began to repeat itself from operator to operator. The study then switched its focus from the gathering of more organizations' policies, to the organization and determination of the importance of what had now been identified in writing, often by multiple operators. Although there is no written standard, a 200+ item line audit would be unwieldy, time-consuming, confusing to those using it, and require a significant amount of effort to customize to a specific facility's needs.

During the content analysis coding phase of the research, a second rater assisted with the objective coding of the large pool of documents utilized in the study. The presence of a second rater contributed to lending face validity to the study. More importantly, the second rater was selected on the basis of their interest and proficiency in content analysis. The second rater's only experience with the workings of a hospitality operation was as a consumer of such products. The thought behind this selection was that, because the primary coder had extensive experience in the hotel industry, and was subsequently quite aware of industry jargon, terminology and situations, a second rater that had less experience with the industry might be less likely to make inappropriate assumptions that erroneously allowed for the collapsing of codes or code structures to make the document smaller and more easy to utilize in a work setting. The second rater engaged in several conversations with the primary rater where he required an explanation of work processes behind some of these descriptions, and often disagreed with the primary coder's attempt to condense certain tasks that seemed or by description sounded similar. Though this might sound frustrating, and often was, the result of having multiple coders allowed for the creation of a coding structure that preserved the specific nuances of varying corporate standards while still collating and organizing all such policies into a digestible and clearly defined structure that the panel of experts could utilize with clarity. After both raters finished the final run of coding, the generic audit that was developing had been reduced to 174 line items.

Sending the generic audit in its full, post-rater structure entitled Initial Audit Draft – Panel Revision (Appendix B) in advance of their meeting was intended to give the individuals on the panel time to view these standards on their own schedule, and in the perspective of their facility or organization, before they joined as a group and could possibly feel uncomfortable speaking up in dissension of the remainder of the group. The researcher sent material out early, and provided clarification as requested on the purpose of their participation and, in some case, the thrust or intent of a specific line item on the audit. Though time availability of the executives participating in the study was a concern and a justification for early release of the generic audit template to the panel members, it was also done to reduce the possibility of groupthink. As it was important with the wide variety of sources for the historical operating data to represent as many perspectives on the presence of varying room service operating procedures as could be located, it was also important to allow individuals to develop and express their perspectives freely and without external influence prior to meeting as a panel.

The panel telephone conference was conducted in an open and comfortable environment with an eye towards inclusion of feedback from all sources in the panel. The generic audit, which had been supplied in an audit style checklist format, was reviewed line item by line item with comments on the importance, wording, categorization, and sub-headings solicited from each panel member. Notes were taken throughout the process.

In several cases, individuals in the panel would create a conversation about what their organization did with regards to a specific policy or procedure in comparison to other organizations. Frequently, in such a scenario, a consensus would arise as to what the spirit of the standard was, and how it could be worded better, or collapsed into another existing passage. In other situations, brand new additions to the audit template were generated from these conversations, often including items that were not extant in the literature or the historical documents. The panel was extremely helpful during this phase in terms of creating firmer structure to the coding schema, and a large and logical series of condensations to the document with which all parties in the panel concurred. This was exactly the reason this study adopted a panel of experts approach. The experts contributed clarity, insight, and real-world workplace logic that would otherwise not have existed with just one or two raters, be they experiences in the hospitality field or not. This kind of multiple step process to checking, reviewing, reorganizing, condensing and collapsing line items lends even more face validity to the process.

At this point, the panel completed their group work, and the researcher took the audit files, the notes, and the written commentary where available from the panel embers and the panel reservists, and revised the audit template to reflect all suggested revisions, deletions and additions. The audit file was then also altered slightly to include a 5-point Likert rating scale in preparation for the next phase, the rating of these items in the scale.

The next version of the audit entitled Panel Rating Instrument (Appendix D), which reflected panel input, was reduced greatly. This version had only 84 line items, each rich with textual description and partial lists of tasks that were included in each line item as a form of examples for the reader or user's benefit. The panel was asked to review the document once again to ensure that revisions discussed in the panel meeting were properly reflected. The panel clearly and fully agreed that this version of the audit was neater, cleaner, easier to understand, and complete enough yet condensed enough to be used as a self-assessment tool by staff conducting such duties in the course of a work shift. The panel then rated all 84 items on the 5-point Likert scale, and the researcher gathered this data, generated average ratings from the nine panel members (seven panel members and two reservists), and re-ordered the audit to reflect the descending order of perceived importance of these items as seen by the panel of experts. As discussed earlier in this study, a 4.0 rating (or higher) was arbitrarily established as the cut-off or the most important items in the 84 item generic audit. This benchmark identified 47 items that were rated as of high importance or of very high importance, and it is these items that answered the first research question regarding what are the most vital elements of a room service operation. These 47 items were also the ones that were utilized to create the audit file that was further tested by the general population in an online survey in the next phase of the research.

This researcher was confident that the 47 line items, identified as Room Service Field Test Directions in Figure 4 by code and in Appendix H Foundational Room Service Practices by text, and created as a result of the various stages of the research outlined previously are the most vital and salient characteristics for a room service operation. The sample population for the online study could consider further support for this assertion from the results of the rating of the same 47 items. Though no statistical analysis was conducted to compare the results of the ratings of the items by the panel of nine versus the sample of 302, it can be noted that not one of the items rated by the online sample was less than 3.16, with three signifying of moderate importance on the 5-point Likert scale utilized. The 47 line items had strong support from two different groups that helped to identify, create, or shape them, as well as strong connections to the literature and the historical documents originally used to source and identify some of them.

During the panel review and discussion phases, the panel made several suggestions for revisions to the original generic audit. The Panel Revisions to Audit Draft (Appendix C) details those revisions. An important suggestion the panel made was to remove the categorical sub-headers in the document. These were the sequential super codes originally developed to assist in the content analysis coding process, drawn originally from the text. The panel noted that there were several passages that were similar in intent across these super codes. They suggested that if these categorical subheaders, which were removed, more similar items from these sections could be collapsed and condensed. For example, the process of employee-guest interaction, and appropriate behaviors associated with that interaction, such as attitude, speech, and use of guest name, appeared in multiple sections. The removal of sub-headers allowed these similar actions to be further grouped regardless of when in the process of executing room service they might appear.

Examination of these 47 line items showed that there were a large number of line items focused on employee behavior, appearance, and guest interaction. Twenty-one of the line items are concerned with how the employee interacts with the guest during the order taking and delivery process, including items such as tone of speech, offering of assistance, requesting of guest preferences, identification of guest and employee name, up selling, uniform quality, and employee service training and preparedness. Another area of focus within these 47 items was preparation for order delivery. Thirteen of the items were focused on standards such as the condition and cleanliness of the delivery equipment, verification of order accuracy, supply stock and shift preparation, and tray delivery and retrieval standards. Only six items were concerned with the actual production of food or beverage products. Two of these were concerned with alcohol service and measurement. There were only four specific food production related policies, focused on standards of recipe standardization and menu item availability. Other remaining areas in this 47 item list included the processing of orders from an accounting perspective, and standards concerned with the verification, processing, and delivery of VIP and amenity orders. It was somewhat surprising that there were so few standards actually related to the consistent production of the core product for room service, the food and beverage items that were ordered by the guest for consumption. The panel ratings for all line items ranged from 3.14 to 4.86. The top four items rated by the panel were all concerned with employee behavior, attitude, and employee-guest interaction.

5.3 Research Question 2

The second research question followed the first closely. Where the first question asked what were the most vital and salient items developed through this multimethod process, the second question was concerned with which of these practices could be seen as foundational to the success of a room service operation. After the panel assisted in the identification of the 47 most important room service policies and standards as previously outlined, these 47 items were posted in an online survey format for review and rating by a sample of 302 individuals with professional hospitality food and beverage experience. The sample was requested to review and rate the 47 items using a 5 point Likert scale in the same fashion as the panel did. Using the same cut off point of a mean rating of 4.0 or higher, this list was further narrowed to 27 items in Foundational Room Service Practices (Appendix H), with mean ratings of those 27 items ranging from 4.01 to 4.32.

Examining these 27 items for patterns revealed a few general categories of policies within this list. Of the 27 items, 12 were concerned with employee attitude, behavior, speech, and guest interaction. Eight of the items were focused on the preparation of the order for delivery, including equipment cleanliness and condition, and order verification. Four of the items were focused on food and beverage production, mainly with the standardized recipe policies for presentation or measurement. The remaining three items were concerned with preparation, verification, and posting of VIP and amenity orders. The top three rated items were all about employee-guest interaction. Whereas the first research question for this study was aimed at determining what the most vital and salient, and generated 47 items considered to be important, the second question was interested in digging even deeper to see what the sample considered to be foundational.

The similarities in the focus on the importance of the employee in the guest service experience between the panel and the sample help to support the assertion that employee behavior is one of the most important practices for ensuring a quality room service operation. In addition, this study found that equipment condition and cleanliness, order preparation and accuracy, and food and beverage production standards are also among the most critical standards for a successful room service operation in a first-class and luxury hotel.

5.4 Research Question 3

The third research question was aimed at the generation of a generic room service audit. This study found it was possible to create an overview of these practices that would reflect the standards and operating practices of most first-class and luxury hotels. The audit, generated through content analysis, was drawn from standards from a wide range of property types, including independent facilities, franchised hotels, resort and nonresort properties, and hotel management companies. The panel members themselves also possessed extensive work experience both in the past and currently at all of these property types. Drawing from as wide a range of participants and content as possible in order to best represent the greater population is a main goal of qualitative analysis.

In order to determine whether or not the audit produced as a result of this analysis was indeed considered to be adequate and appropriate for use in an actual operation, a field test of the audit was conducted. A luxury facility on the Las Vegas Strip was solicited to allow the use of the Room Service Field Test Directions (Figure 4) in their room service operation. As discussed in the results section, the field testing was very successful. All staff members utilized the audit for three or more separate shifts, and employee and management feedback were solicited throughout the process. In general, the staff and the management of the operation found the audit more than adequate. They found it easy to use and clear in its language and layout. A strong point supporting the viability of this tool can be found in the way in which the staff, with no real instructions to do so, immediately took the audit and began to customize it for use at their property by adding or altering information within the text of the line item to make it represent the established and current practices of the facility. It would not be possible to make an audit that specifically represented the differentiating service standards of every type or brand of hotel in every geographic location. So, the goal of this study was to utilize existing documents and professional, expert feedback to take detailed passages on standard operating procedure and make them less specific and more general. The experiences of the researcher observing the audit being utilized, and the feedback from the staff and management that utilized the document in a real work setting supported this audit as a viable tool for properties with a room service operation regardless of any specific or unique standard they may have adopted over time.

5.5 Research Question 4

The last research question was related to the third question. The third question was interested in determining whether a generic audit that was universally representative of functions that needed to be performed in order to have a successful room service operation.

This study determined it was possible and developed one. The fourth question wanted to further advance the notion of whether an audit generated for such purposes might have any noteworthy differences between property types. It was decided that the online survey instrument would seek out differences by facility type, and asked respondents about what kind of facilities they worked at in their employment history.

The types of hotels investigated were broad. They included resort vs. non-resort, large versus small, the presence vs. the absence of a free-standing room service department, and franchise vs. independent properties. Hypotheses were developed for investigating the differences in the perception of importance of the established general standards across hotel types. As reviewed in the results section, there were no significant differences in the perception of importance of room service policies between large and small hotels, resort and non-resort hotels, and hotels with free-standing room service departments and those with room service incorporated into an existing outlet. There was a significant difference in the perceived importance of these standards between a franchised hotel and an independent property. Examination of the difference showed that these two groups rated more than half of the items differently. What was interesting is that in every difference in rating, the franchised hotel rated the item as more important than the independent operator.

This was somewhat of a surprise, as the researcher expected to find that the franchised hotels were more focused on measurable standards such as delivery time, strict standard adherence, and order accuracy. The researcher also expected to find that the independent hotels were more focused on the importance of employee-guest interaction and the creation of a quality, memorable experiences. This was not the case. Rather, the independent hotels seemed to be far less interested in standards overall than their franchised counterparts.

Franchised hotels leverage their brand identity through the creation of various standards and the marketing of the position that those standards help to reinforce through consistency. Independent hotels do not necessarily have a corporate office that develops and directs corporate policies and standards. A franchise corporate office acts as a monitoring tool for its members, devising and communicating standards, and utilizing quality assurance programs and inspections to verify and reinforce these standards. There is often an entire department in a franchise organization whose only job is to develop standards and see them implemented consistently. It is proposed that the absence of such a department at a corporate level, removed from the daily operational idiosyncrasies, is what generates these differences in perceived importance of certain standards. It is not that the independent hotel does not know or care what is important, but rather that there is no specific entity in their organization that makes standards a sole focus. An independent hotel also needs to position itself as a unique facility, and may not see the importance of specific standards as they relate to brand image or consistency across locations, as might a corporate office in a franchise organization.

There was no indication that a reduction in the perceived importance of a set of standards would require that the audit itself be structured or designed differently. Rather, an independent hotel could utilize the same generic audit, and simply customize it to more closely reflect which practices are important and to be measured. It should be noted that an independent hotel could likely benefit greatly from access to the standards franchise hotels utilize and are reflected in this study's generic audit, in effect getting corporate franchise support and input while still retaining its independence. With regards to the final research question, there are, from both quantitative and qualitative assessment, no significant differences of a structural nature among hotel types with regards to the creation, design, or content of a generic audit.

5.6 Academic Significance

This study has academic significance. This research contributed to the academic body of knowledge on room service operations. In the literature, there was very little written about room service. This work generated a robust and fully detailed overview of the function and purpose of, and the means by which, hotels conduct luxury room service operations. By synthesizing the literature in hospitality textbooks, trade magazine articles on room service, and the few academic studies that exist in this field, this study has contributed to the body of knowledge on the workings of a room service operation. With regards to the literature, it was discovered that very few of the trends or policies discussed in the trade magazine articles and in some of the textbooks were actually an issue of perceived importance once investigated. For instance, there were several articles on the importance of menu engineering through locally sourced products, regionally specific products, or price and consumption analysis. However, none of the historical documents and none of the panel's input indicated this to be true. There is nothing in this audit about menu engineering. Similarly, there is nothing in the audit that agreed with the assertion that elevators should or could be reengineered to become floating kitchens in order to better serve room service patrons. From this perspective, the study was an important contribution to academic significance, as it reminded us that what is often written in trade journals, the popular press, or even textbooks is often an opinion or a 116 snapshot of a trend, and has not necessarily been verified snapshot of a trend, and has not necessarily been verified thorough robust analysis or peer review.

The results of the study further enhanced the quality control and constant quality improvement literature by bridging the theory with an applied, research based tool. The same was true for the operational auditing theoretical process as discussed in the literature review. The results provided further framework for implementation of the operational auditing processes.

This study also extended the works of Moreo, Sammons and Savage (1997), and Woods, Moreo and Sammons (2005). It added another hotel department to those that have been completed using similar methods of distilling generic procedures from existing historical documents. It also extended their work methodologically, by adding the panel of expert verification process and survey components. Both of these extensions contributed by creating a methodological model that focuses further on face validity, and the generation of more intimate knowledge through a robust and rigorous 5-step multimethod data generation process.

5.7 Practical Significance

The practical significance of this study was in its application in the industry. This study provided potentially beneficial contributions to hotel corporations, independent properties, and individual employees of a hotel. For a corporation, be it a franchise organization or a hotel management company, this information can be added to the firm's corporate quality assurance efforts. If the firm does not already have such an instrument, it can easily be customized and adopted to create an assessment tool for property visits or analysis of service deficiencies. Because it was drawn from a wide range of properties, a firm could also utilize it as a form of competitive analysis. It can compare its standards to that of the larger population and see where the hotel might improve its offerings through clear communication of desired standards. For smaller firms, an additional benefit might be that of time or money savings, as smaller firms may not have the manpower to dedicate to the lengthy and time consuming task of detailing standards of operations from various entities in the industry. An independent property can benefit in the same way. It can develop a set of standards that borrows from established practices in order to remain competitive in terms of consumer expectations. It can utilize the standards from the industry as a whole to determine where it might increase the level of standards for a specific policy in order to generate a competitive advantage through product differentiation. As with a smaller corporate firm, funding might not be readily available to dedicate to the effort required to develop such a comprehensive list of standards internally. A hotel management company could also benefit from the application of a generic audit. Management companies often manage multiple brands. The ability of this tool to reflect industry standards absent of brand, and the ability of the generic template to be readily and easily customized to fit a specific brand can save a lot of time for management firms taking over new contracts. The audit can also be fashioned to function as a mystery shopper tool, internally or externally by any organization that sees the benefit of internal self-assessment.

There are also benefits for individual employees in a first-class and luxury hotel in the application of the generic audit. As experienced during the field test, employees quickly gravitated towards the audit as a form of self-assessment. With a concise implementation plan from management, this audit can empower employees to take more responsibility in assuring established standards are met before being directed to do so by management. In this way, it can be used as a tool for continuous quality improvement. The flexibility of the document also allows it be used as an incentive tool for employee compliance to standards. As was done by the management in the field test, the document can be divided into specific and logical subsections that are then distributed directly to the staff members responsible for these duties. It also has value as a training tool, to communicate a high level of detail and explanation behind each of the standards in place. If management emphasizes the importance of the tool, and the staff is permitted to customize the tool in a way that accurately reflects the way in which the policy is implemented, the audit can be a strong focusing tool that compels the staff and management to use it regularly and keep it current. This, in turn, can focus the entire organization towards real and measurable goals that contribute to matching customer expectation to customer perception, thereby creating higher perceived quality between guest and employee.

5.8 Findings

This study generated several interesting findings. It assisted in the creation of a robust and representative generic RS audit which is customizable and can be selfadministered. As a result of the multiple steps in the research, this study identified foundational RS practices that are high in reliability, and face and content validity.

From the quantitative analysis, it was determined that there were no significant differences in the perception of importance of identified RS practices across three groups of operators; large vs. small hotels, resort vs. non-resort hotels, and hotels with freestanding RS departments vs. those without. There were significant differences in perceptions of importance of RS practices between independent and franchise hotel operators. 21 items perceived differently between these two groups. Independent operators saw all 21 as less important. The main differences were in the areas of employee behavior, preparation of the order for delivery, food and beverage production, and the handling of VIP and amenity orders.

As mentioned previously, it is proposed that the main differences in perceptions of importance of key RS operating practices between franchised and corporate hotels is due to the absence of a monitoring and inspection effort in an independent property. Such efforts are commonly found in franchise hotels at both a corporate and a unit level, but generally non-existent in independent facilities. The lack of difference in perceptions of importance of key RS operating practices in the other hypotheses in the study bears discussion as well. The size of the hotel had no apparent effect on the perceived importance of key RS practices. It was originally considered that a larger facility might have more resources or more employees for the performance of specific tasks. Perhaps this in an indicator that most facilities streamline their operations, and that equipment and production processes are enhanced with technologies that allow smaller facilities to handle variances in volume without sacrificing service quality. Resorts were compared to non-resort facilities. Resorts tend to offer a wider array of personalized services than their non-resort counterparts, and that resort properties were generally more focused on creating personalized and memorable guest experiences. It is the belief of this researcher that, while this may be true, the results indicate that, in terms of operating procedures, there are no specific RS differences germane to a resort. The foundational practices indicate that employee-guest interaction is an importance aspect of RS practice regardless of hotel type. This study also examined the perceived difference between a hotel with a free-standing RS department and one with no free-standing RS department. If the demand exists for room service in a facility, but the physical layout does not provide production space for RS, then it stands to reason that RS was not originally planned for,` but rather grew out of customer-driven or brand-driven necessity. If this is the case with those facilities with no free-standing RS departments, then it would stand to reason that these facilities would emulate an established set of policies, and that any customizations to those policies driven by logistical or space necessity would not be great enough to alter the importance of the standard RS practices. All of the three hypotheses that were not rejected share a common methodological approach. The collection of historical documents, the selection of members for a panel of experts, and the online respondents were all purposive samples, aimed at creating a representation of the industry as a whole. To fail to find significant differences in these hypotheses may be a result of the data collected being general enough to apply to various segments of the hotel industry, one the main goals of this study.

As with any exploratory study, the literature served to guide the study's efforts. The literature helped to develop an overview of RS trends, challenges, history, and structure. It also assisted in the development of the hypotheses for the study. Literature in the areas of service quality and operational audits helped to create a framework for understanding the importance of the creation of an operating model for RS operations to follow. This study discovered that there was very little linkage between the literature in RS and the results of this study. A majority of the practices identified in the literature were not perceived to be important by RS operators in the study. Herein lays one of the great values of this study.

Though literature can help guide us through a process of discovery, in a constantly changing, trend-driven business such as hospitality, it cannot necessarily accurately reflect the industry as a whole. Rather, as seen in this study, it can only serve as a snapshot in time. Menu engineering and marketing were both seen in the literature as important phenomena that would greatly impact a RS operation. In this study, neither of these items appeared anywhere on the list of important practices, even early in the development of the overview of all practices. Neither the historical documents, nor the panel of experts identified marketing or menu engineering as important. Even information extracted from academic hospitality textbooks was determined to be inaccurate. One of the six major areas of RS operations as identified in the text was routing, the process of forwarding an order from the RS dispatcher to various production facilities in the operation. Routing was not noted in the historical documents, nor discussed by the panel. It should be considered that routing was likely seen as important prior to the introduction of technology that made the dispersion of production orders a mechanical function as opposed to a human one. Once technology was introduced into this process, the potential of human error was reduced significantly, and the routing process was no longer seen as a procedure that required heavy oversight.

Literature that suggested the importance of operational auditing was supported by this study, as was the development of a measurable instrument aimed at improving service quality. This study built on the framework provided by the theoretical literature by adding a new step to the process of hospitality operational auditing; the verification of the audit content, structure, and application by professionals in the field. In general, this study determined that literature in an exploratory study should be used to guide the process, but also that it can be inaccurate, and should be looked at with a critical eye, especially in cases where the literature is dated, and the subject material is customer preference and trend-driven, and tends to evolve over time.

It was of interest that the resultant audit and the list of foundational practices had so little emphasis on standards regarding the core product, the food and beverage items prepared for delivery and consumption to a guest room. The few standards that existed were primarily about portion size and measurement, and recipe standardization, and were limited to only two of each for beverage and food production. This is only 4 items out of 85, a relatively low amount of attention to be paid to one of the few tangible elements of this service experience. Perhaps further investigation could reveal that culinary standards are contained within another set of policies and practices, more often the responsibility of a culinary department directly. Perhaps the large variety of menu items, cooking styles, and methods preclude the ability to easily standardize the actual production processes, but it seems that an area of opportunity exists if there are no firms paying greater attention to the consistency and measurability of its food and beverage offerings.

The areas identified by the study as foundational practices are telling. The final list of 27 foundational practices can be further grouped into topic areas that should become the focus of hotel RS departments. These areas include employee-guest interaction, product delivery standards, order verification, and food and beverage production. Even if facilities consider the process of customizing and utilizing an audit to improve service quality too costly or time consuming, they can benefit from concentrating on the development of training or measurement tools in these areas.

5.9 Limitations

Generally, this study was rich in detail and purpose, and careful about its rigor in designing the methodology and analyzing the data. All studies have limitations. One limitation of this study was the field test. Only one property was utilized for field testing. Though the field testing was smooth, successful, and generated excellent feedback, it was conducted at a very large corporate facility with detailed and established standards. In light of the findings that independent hotels seemed to rate many of the standards on the audit as less important than their franchised counterparts; it would have been more effective to see how the audit was perceived and utilized by a staff in an independent hotel. Additionally, a series of visits to such an independent facility might have contributed a greater understanding as to why these hotels seem to rate certain policies as less important.

Another limitation in this study was the lack of food production standards. Very little of the content regarding operating procedures that was generated had anything at all to do with the actual production standards behind the general act of cooking dishes for a room service operation. Perhaps the study should have included culinary departments in the collection process, or specifically requested culinary production materials or input from the firms that supplied historical documents. It is puzzling to consider that a service operation that delivers food and beverage does not seem to have a focus on the actual production standards behind the creation of a dish. It could be that some of these standards exist in a culinary document used by the kitchen staff and not the service staff. Though it is quite likely that food production standards are greatly affected by the foodstuff being cooked, much more so than it affects the preparation for the delivery of that product, it still seems like a little more information on cook and hold temperatures, menu product engineering, and specific cooking methods for dishes common to a room service operation would have benefitted this study in terms of generating an even more detailed overview of the room service process in a first-class and luxury hotel.

5.10 Future Research

A future research suggestion would include conducting similar studies in other underexplored areas of hotel services and operations. Based on the previous discussion, culinary would be a recommended area of investigation, especially considering that it is a part of several other operations, including banquets and restaurants. Other areas might include information technology, sales and marketing, facilities management, recreation, and casino services, to name just a few. A large goal for this researcher would be to continue this method of research to fully explore all other typical hotel departments and operations. Conducting such studies and combining them into some form of a handbook of hotel operations that was beneficial to both operator and researcher would be the ultimate focus of such a large, multifaceted research endeavor.

Using the data generated in this study as a foundation, another potential future research project would be to further investigate these policies in detail. Can these most important policies be further defined and tested with the goal being the identification of critical success factors for room service? Is there relative weight to the importance of one set of policies versus another? It seems that, using the constructs developed in this study, further investigation could be made using exploratory cluster analysis, followed by factor analysis if the results from the cluster analysis were significant. Determining not just what policy was important, but which policy contributed the most to profitability or guest satisfaction for example, could produce results that would help both the industry and academia in determining where to focus resources and efforts to maximize the return on their operations.

5.11 Conclusion

This study used a multimethod approach to investigate room service operating policies. Using a five-step multimethod approach that included content analysis, panel of expert review, field testing, and quantitative analysis, it generated several noteworthy findings. It developed a comprehensive list of policies and procedures currently in place in room service operations in first-class and luxury hotels. It developed a robust coding schema for use in future similar studies. It verified the inclusion of line items in a generic audit through both panel of experts review and quantitative ratings. It developed a full list of room service policies that was generically representative of common practices, while still being customizable to the needs of a specific property or brand. It identified the most important room service practices within the industry at several levels. It created a selfassessment audit tool for use in hotel room service operations that is usable by line staff, customizable by management, and measurable by all parties. It tested this tool in an actual operation to determine if it was appropriate for the uses for which it was originally designed. It investigated differences in perceived importance across several different types of facilities in order to ensure a robust and representative tool. The audit that was created as a result of this study was strong, comprehensive, clear, and flexible, and can be readily used by a wide array of scholars and practitioners to improve operations or increase understanding.

6. ACKNOWLEDGEMENTS

William F. Harrah College of Hotel Administration at the University of Nevada, Las Vegas, provided the funding for this research.

I would like to thank my chairperson and committee members for their dedication and commitment to this research. Without their input, collectively and as individuals, I would not have been able to complete this study. Their knowledge and guidance will serve to direct my scholarly efforts for the rest of my career, and for that I am grateful.

I would like to thank all of my professors from my time here at UNLV. Without their teaching, I would lack the tools to undertake this project, as well as the insight and confidence to pursue this research.

I would like to thank my professional associates in the panel of experts, and my colleagues involved in the coding process. Without their participation and input, this study would lack the accuracy, robustness, and completeness that a dissertation should contain.

I would like to thank my editor. Her expertise in editing made it easier to concentrate on the content of my work, and her collaboration on the tables and figures for the study saved me from insanity.

I would like to thank the administrative and support staff at UNLV. Without their patience and direction, this study could not be completed.

I am grateful and proud to have had the opportunity to interact with all of them, and am hopeful that there will be more opportunities to do so in the future.

REFERENCES

1. Augustyn, M., & Ho, K. (1998). Service quality and tourism. Journal of Travel Research. 37, 71-75.

2. Berg, B. (2001). Qualitative research methods for the social sciences. Boston, MA: Allyn and Bacon.

3. Bernstein, C. (1989). The room-service nightmare. Nation's Restaurant News, 23(11), F3.

4. Brewer, J., & Hunter, A. (1989). Multimethod research: A synthesis of styles. Newbury Park, CA: Sage Publications, Inc.

5. Brooks, K. (1979). Delphi technique: Expanding applications. North Central Association Quarterly, 53, 377-385.

6. Buick, I. (1998). A preliminary study of the operational aspects of providing room service to hotel guests in Edinburgh's principal hotels. International Journal of Hospitality Management, 17(1), 49.
7. Cichy, R., & Wise, P. (1990). Managing service in food and beverage operations. Lansing, MI: Educational Institute of the American Hotel & Motel Association.
8. Churchill, D. (2006). Are you being served? Business Travel World, 37-39.
9. Comen, T. (1989). Making quality assurance work for you. Cornell Hotel & Restaurant Administration Quarterly, 30(3), 23.
10. Davis, B., & Stone, S. (1985). Food and beverage management. Oxford, England: Reed International Books.
11. Easterby-Smith, M., Thorpe, R., & Lowe. A. (2004). Management research: An introduction. London, England: Sage Publications, Inc.
12. Eaton, D. (2004). Legal consequences of sexual harassment of staff by hotel guests. The Cornell Hotel and Restaurant Administration Quarterly, 45(4), 347-361.
13. Finn, M. (1977). Room service: break-even convenience or profit center? Lodging Hospitality, 35(12), 39-40.
14. Flesher, D., & Siewart, S. (1982). Independent auditor's guide to operational auditing. New York, NY: John Wiley & Sons, Inc.
15. Gabor, A. (1990). The man who discovered quality. New York, NY: Times Books, Random House.
16. Glover, W., Morrison, R., & Briggs, A. (1984). Making quality count: Boca Raton's approach to quality assurance. Cornell Hotel & Restaurant Administration Quarterly, 25(1), 39.
17. Gronroos, C. (1984). A service quality model and its marketing implications. European Journal of Marketing. 18(4), 36-44.

18. Hanson, B. (1986). Hotel food service: where's the profit? The Cornell Hotel and Restaurant Administration Quarterly, 25(2), 93-97.

19. Hair, J., Black, W., Babin. B., Anderson, R., & Tatham, R. (2006). Multivariate data analysis. Upper Saddle River, NJ: Pearson Prentice Hall.

20. Hemp, P. (2002). My week as a room service waiter at the Ritz. Harvard Business Review, 80(6), 50.

21. Hubsch, A. (1966). Hotel food and beverage management. The Cornell Hotel and Restaurant Administration Quarterly, 7(1).

22. Johnson, B., & Onwuegbuzie, A. (2004). Mixed-methods research: A research paradigm whose time has come. Educational Researcher, 33(7), 14–26.

23. Johnson, R., & Turner, L. A. (2003). Handbook of mixed methods in social and behavioral research. Thousand Oaks, CA: Sage Publications, Inc.

24. Kandampully, J. (2007). Services management. Upper Saddle River, NJ: Pearson Prentice Hall.

25. Karipidis, P., Athanassiadis, K., Aggelopoulos, S., & Giompliakis, E. (2009). Factors affecting the adoption of quality assurance systems in small food enterprises. Food Control. 20, 93-98.

26. Kassarijian, H. (1977). Content analysis in consumer research. The Journal of Consumer Research, 4(1), 8-18.

27. Keen, D. (1994). Can you top this? How Six Hong Kong hotels run room service. Hotels, 28(4), 71.

28. Keppel, G., & Wickens, T. (2004). Design and analysis: A researcher's handbook. Upper Saddle River, NJ: Pearson Prentice Hall.

29. King, C. (1984). Service-oriented quality control. Cornell Hotel & Restaurant Administration Quarterly, 25(3), 92.

30. King, P. (2006). The road to room service. Foodservice Director, 19(8), 24-26.

31. Lau Pei, M., Khatibi Akbar, A., & Yong Gun Fie, D. (2006). Measuring service quality and customer satisfaction of the hotels in Malaysia: Malaysian, Asian and nonAsian hotel guests. Journal of Hospitality & Tourism Management, 13(2), 144- 160.

32. Linstone, H. A. (1978). Handbook of Futures Research. Westport, CT: Greenwood Press.

33. Locke, K. (2003). Grounded theory in management research. London, England: Sage Publications.

34. Marshall, C., & Rossman, G. (1999). Designing qualitative research. Thousand Oaks, CA: Sage Publications, Inc.

35. Mehmetoglu, M., & Altinay, L. (2005). Examination of grounded theory analysis with an application to hospitality research. International Journal of Hospitality Management. 25(1), 12-33.

36. Merriam, S. (2002). Qualitative research in practice. San Francisco, CA: Jossey-Bass.

37. Miller, B. (2005). Hospitals go gourmet with room service for patients. South Florida Business Journal. July 29, 2005.

38. Mohsin, A. (2007). Assessing service down under: A case of Hamilton, New Zealand. International Journal of Contemporary Hospitality Management, 19(4), 296-308.

39. Moreo, P., Sammons, G., & Savage, K. (1997). Hotel operational audit: A questionnaire approach. Bottomline, 12(6), 26.

40. Moreo, P., & Savage, K. (1990). Incorporation of operational audits in hotel front office courses. Hospitality Research Journal, 14(2), 243.

41. Morgan, D. (1997). Focus groups as qualitative research. London, England: Sage Publications, Inc.

42. Muhr, T. (2005). Atlas/ti 5 – Software for qualitative data analysis. [Computer software and manual]. Retrieved September 18, 2010 from www.atlasti.com.

43. Narangajavana, Y., & Hu, B. (2008). The relationship between the hotel rating system, service quality improvement, and hotel performance changes: A canonical analysis of hotels in Thailand. Journal of Quality Assurance in Hospitality & Tourism, 9(1), 34-56.

44. Ninemeier, J., & Purdue, J. (2008). Discovering hospitality and tourism: The world's greatest industry. Upper Saddle River, NJ: Pearson Prentice-Hall.

45. Pfeffer, J. (1982). Organization and organization theory. Marshfield, MA: Pitman Publishing, Inc.

46. Reider, H. R. (1994). The complete guide to operational auditing. New York, NY: John Wiley & Sons, Inc.

47. Rutherford, G. (1990). Hotel management and operations. New York, NY: Van Nostrand Reinhold.

48. Seidman, I. (2006). Interviewing as qualitative research. New York: Teacher's College Press.

49. Serlen, B. (2003). Hoteliers see uptick in in-room dining. Business Travel News, 20(5), 14.

50. Stellin, S. (2007, January 30). Rewriting the room service menu. The New York Times, pp. A0, B0.

51. Thornhill, W. (1981). Complete handbook of operational and management auditing. Englewood Cliffs, NJ: Prentice-Hall, Inc.

52. Turk, D. (1979). Room Service: No longer a necessary evil, Lodging Hospitality, March, 1979, 26-28.

53. Vallen, G.K., & Vallen, J.J. (2008). Check-in check-out: Managing hotel operations. (8 ed.). Englewood Cliffs, NJ: Prentice Hall.

54. Weber, K. (1997). Assessment of tourist satisfaction using the expectancydisconfirmation theory. Pacific Tourism Review. 1, 35-35.

55. Willborn, W. (1986). Quality assurance audits and hotel management. The Services Industries Journal. 6(3), 293-308.

56. Wolf, C. (1992). Room Service blues. Lodging Hospitality, 48(12), 55-56.

57. Wood, D., Moreo, P., & Sammons, G. (2005). Hotel housekeeping operational audit: A questionnaire approach. International Journal of Hospitality & Tourism Administration, 6(3), 1-10.

58. Yousuf, M. (2007). Using experts' opinions through Delphi technique. Practical Assessment, Research & Evaluation. 12(4).

4

IMPROVING PATIENT SATISFACTION IN A HOSPITAL FOODSERVICE SYSTEM USING LOW-COST INTERVENTIONS: DETERMINING WHETHER A ROOM SERVICE SYSTEM IS THE NEXT STEP

Vanessa A. Theurer

Utah State University

ABSTRACT

A survey was used to assess patient foodservice satisfaction at an acute care hospital. Results were then used to identify the lowest scoring areas of foodservice. Low-cost interventions were implemented to address the areas of needed improvement. Patient satisfaction was reassessed by re-issuing surveys. Group T-tests were used to compare the results of the baseline and follow-up surveys (P, 0.05). Low-cost interventions were not shown to be effective in significantly improving patient foodservice satisfaction. Findings suggest that a higher-cost investment in a room service system maybe the next step in order for the foodservice operation to meet patient expectations.

KEYWORDS

Hospital foodservice/room service, low-cost foodservice interventions, patient foodservice satisfaction

PART I

1. STATEMENT OF PROBLEM

Hospital food and nutrition services play an important role in patient recovery and well-being. Foodservice quality can also influence patients' satisfaction with their overall hospital experience (McLymont, Sharon, & Stell, 2003; Williams, Virtue, & Adkins, 1998). With health care industry competition on the rise, many hospital foodservice operations are looking for ways to improve patient satisfaction. This study will investigate patient foodservice satisfaction at St. Mark's Hospital (STMH), a 276- bed teaching hospital which primarily serves the urban community of the Salt Lake Valley in Salt Lake City, Utah.

STMH is part of the Hospital Corporation of American (HCA) and is a highly respected Mountain Star facility. HCA is a Nashville-based hospital company with 164 hospitals in 20 states (Hospital Corporation of America, 2011). HCA conducts a quarterly, company-wide Gallup poll that surveys post-discharge patients on a variety of factors affecting their overall hospital stay experience and satisfaction. Gallup is a global organization that provides statistical research services (Gallup Organization, 2011). For several years, STMH has consistently scored below the company average in satisfaction in overall quality of foodservice. The single question patients are asked during the telephone interview is, "Were you satisfied with the overall quality of foodservice?" The survey poll does not provide any further details as to why post-discharge patients are dissatisfied.

For years, the STMH foodservice department has also distributed an internal questionnaire of its own on the back of the patient menu. It consists of six “yes or no” questions, with little room to add comments. The questions address meal taste, temperature, appearance, reliability, service and overall satisfaction. According to foodservice department records, the questionnaire has a very low participation rate, with less than 3% of patients returning a completed survey. For those that are completed, the data has not historically been analyzed or recorded in a standardized way.

The food at STMH is prepared on-site in a centralized kitchen. Hot thermal retention trays are then assembled in a round tray line using heated bases and transported via elevator in carts. The foodservice staff delivers trays directly to patients in their rooms. The foodservice department uses a selective 7-day menu cycle. Diet order-entry is computerized, and menus are manually prepared, distributed, and tabulated once daily.

STMH has a limited budget and foodservice interventions must be made using current resources. In order to control costs while also improving the quality of foodservice, it is essential to isolate dimensions of foodservice that are in most urgent need of improved satisfaction and to identify precisely which effort will reap the largest benefits. This study will investigate the dimensions of patient foodservice satisfaction at STMH in an effort to provide cost effecti to provide cost effective means of improving the patient satisfaction score.

1.1 Problem

Perceived foodservice quality contributes to patient recovery and overall satisfaction of hospital stay (McLymont et al., 2003; Williams et al., 1998). However, in recent company-wide polls, post-discharge patients have scored STMH as below average for satisfaction in overall quality of foodservice. In addition to this, the STMH foodservice department currently uses an ineffective internal patient survey tool. The survey is rarely completed by patients (,3% participation rate) and data collected from the survey is not analyzed.

1.2 Purpose and Objectives

Purpose

The purpose of this study is to determine implications for improving foodservice satisfaction at St. Mark's Hospital using a validated tool.

Objectives

To improve foodservice satisfaction at STMH the objectives include:

1) Identify the food, service, and patient variables that contribute to foodservice dissatisfaction utilizing a validated patient survey.
2) Plan and implement cost-effective changes in the foodservice based on survey results.
3) Monitor foodservice satisf action scores.
4) Determine and recommend further changes on to improve patient foodservice satisfaction to administrators.

2. REVIEW OF LITERATURE

2.1 The Role of Hospital Foodservice

Adequate nutrition intake is an important part of healing the hospital patient. In general, undernutrition is associated with loss of muscle strength and impaired immune function which can lead to an increase in complication rates, infection rates, and mortality (Giner, Laviano, Meguid, & Gleason, 1996; Johansen, Kondrop, & Plum,

2004). Promoting optimal nutritional status through quality hospital foodservices can lead to a faster recovery and decreased length of hospital stay which can have a large impact on hospital costs (Giner et al., 1996; Johansen et al., 2004).

Health care in the United States is the largest service industry in the world and it exists in a dynamic, competitive environment, where the need to provide quality services with limited resources is vital (Drain, 2001; Kizer, 2001; Fallon, Gurr, Hannan-Jones, & Bauer, 2008). Health care institutions must monitor and address patient satisfaction in order to remain viable. Foodservice quality is significantly correlated with overall patient satisfaction (Sheehan-Smith, 2006). Therefore, it is not surprising that many hospital foodservice organizations are changing to be more focused on patient care in an effort to boost patient satisfaction and control costs (Buzalka, 2008; Drain, 2001; Urden, 2002). In the highly aggressive health care industry, hospital foodservice is poised to play an important role in gaining the market share edge.

2.2 Customer-Oriented Hospital Foodservices

The U.S. health care system is being actively reshaped by the expectations of consumers. Heightened focus on hospital stay satisfaction is the result of many numerous interrelated issues, such as the aging of "baby boomers", greater prevalence of chronic conditions, the expansion of biomedical scientific knowledge and technology, and disproportionate health care cost increases (Kizer, 2001). This new era of health care consumerism combined with escalating financial pressure presents massive challenges for hospital administrators. Stiff competition has forced administrators to place greater emphasis on customer-oriented service and cost-effective quality improvement (Goehring, 2002; Kizer, 2001).

The link between a customer-oriented service culture and patient satisfaction is becoming clear. One of the most important principles in customer-oriented service is creating an environment that meets or exceeds patient expectations (Fottler, Ford, Roberts, & Ford, 2000). Establishing a service culture that meets patient expectations requires a total commitment to customer-oriented service. According to a trade journal article by Goehring (2002), hiring and promotion practices should focus on employees who are people-oriented and able to achieve customer service goals.

Another way to establish a service culture is through a foundation of ongoing customer service training. Training can help define service excellence in a hospital foodservice department and throughout the organization (Goehring, 2002; SheehanSmith, 2006). Improving patient satisfaction is also linked with satisfied employees (Goehring, 2002; Norton, 2008). As employees find purpose and meaning in their work, they are motivated to see how their actions contribute to better patient care. Following strategies to improve customer-oriented service will ultimately enhance patient satisfaction (Goehring, 2002; Norton, 2008).

2.3 Meeting Patient Foodservice Expectations

It is well recognized that food and other aspects of foodservice delivery are important elements of the patients' overall perception of the hospital experience. The greater patients' expectations are met, the more satisfied they seem to be (Lau & Gregoire, 1998; Fottler et al., 2001). Therefore, provision of foodservices that not only meet but exceed the expectations of the patient should be considered essential for quality hospital foodservices (Fallon et al., 2008).

2.4 Dimensions of Foodservice

A thorough knowledge of the varied dimensions of hospital foodservice satisfaction is needed to understand patient expectations and promote the health and nutritional status of patients (Wright, Connelly, & Capra, 2006). In a survey questionnaire designed study by Dubé, Trudeau, and Belanger (1994), seven dimensions were identified that represent patients' perceptions of foodservice, including: food quality, service timeliness, service reliability, food temperature, attitude of staff who deliver menus, attitude of staff who deliver meals, and customization. Through 132 hospitalized patient surveys, Dubé et al. (1994) found that food quality was the best predictor of patient satisfaction, followed by customization and attitude of the staff that deliver menus. Other studies by Lau and Gregoire (1998) and Williams et al. (1998) also confirmed food quality as the key factor in patient satisfaction.

Food Quality. The perception of food quality can depend on several different attributes, including meal taste, variety, flavor, the texture of meat and vegetables, the perception of choosing a healthy meal (Wright et al., 2006). Menu changes implemented to improve food quality must, therefore, address a wide range of influences. Each hospital foodservice institution is unique and interventions must be customized to the specific patient population's needs and perceptions (Wright et al., 2006).

In an article by Vozenilek (1999), the successful improvement of one hospital foodservice institution in Dallas, Texas was described. As a registered dietitian and the director of Nutrition Services, Mary Kimbrough led an effort to completely overhaul the patient menu and hospital food quality. Some food quality interventions included adding a variety of fresh fruits, vegetables, and salads to the daily menu; offering more trendy entrees (ie. gnocchi tartlets) along with more familiar comfort food; adding flavors enhancers to sauces, like roasted garlic and fresh herbs; enhancing meat flavors with salsas, vegetable slaws, and fruit purees; enhancing vegetable flavors with fresh herbs, stocks, and aromatics; adding plant-based entrees to please vegetarian customers; and selecting lean meats and preparing them using healthy cooking techniques. These changes yields improved patient satisfaction as well as several foodservice awards (Vozenilek, 1999).

In a 585-bed hospital in Canada, food quality improvements were made on a smaller scale but still focused on specific patient preferences (Watters et al., 2006). Menu changes included adding a weekly seasonal fruit rotation and offering more soup. Changes were also made to the cardiac diet to improve consistency with heart health guidelines and to include more fresh fruit and fish in the menu (Watters et al., 2006).

A validated survey tool should be used to measure and address patient perceptions of food quality and overall foodservice satisfaction. The literature (Dubé et al., 1994; Urden, 2002; Lau & Gregoire, 1998; Williams et al. 1998, Watters et al., 2006) further emphasizes the need for a patient survey to address and thoroughly differentiate the main dimensions of foodservices.

2.5 Patient Surveys

Patients are essential sources of data regarding service function. In most health care institutions, quality-control programs have been implemented and patient satisfaction surveys have been routinely administered (Avis, 1995). However, standard hospital patient satisfaction surveys usually only ask a few general questions about foodservice. Global statements often used in these standard internal surveys (e.g. a single rating of "food quality") provide insufficient information to allow managers to adapt foodservices to suit patient preferences and expectations (Wright et al., 2006). Draper, Cohen, and Buchan (2001) outlined the difficulty of interpreting the results of health care satisfaction surveys when the survey items are not specific enough or appropriately focused, which leads to poor application of the results to service improvement efforts and limits their broader applications to policy.

In a study by Capra, Wright, Sardie, Bauer, and Askew (2005), researchers worked to design a valid, comprehensive and reliable questionnaire to measure patient satisfaction with acute care hospital foodservices. Building on the research of many previous patient survey studies, such as Dubé et al. (1994), The Acute Care Hospital Foodservice Patient Satisfaction Questionnaire (ACHFPSQ) was designed. It was then administered to a convenience sample of 2347 acute care hospital inpatients and postdischarge patients from three hospitals. Results of this questionnaire revealed four main dimensions or categories of foodservice: food quality, meal service quality, staff/service issues, and the physical environment. The study also indicated that the survey was an accurate, reliable measure of patient foodservice satisfaction and allowed for collection of detailed information about attributes within the four main categories of foodservice. Capra et al. (2005) asserted that the ACHFPSQ provides a tool for the continuous assessment of foodservice quality and satisfaction over time in a variety of acute care settings.

A study by Fallon et al. (2008) demonstrated the effectiveness of the ACHFPSQ in a 440-bed acute care private hospital. Surveys were administered to 551 patients over a 3-year period. The study concluded the tool was sensitive enough to detect specific food quality issues from which targeted interventions for quality improvement activities were then focused. The study also found the tool was practical to administer within existing department resources with minimal disruption to work practices.

Similarly, Porter and Cant (2009) found the ACHFPSQ effective at explaining the dimensions of patients' perceptions of foodservice quality which underlie satisfaction. The benefits of using the ACHFPSQ as a total quality management tool was highlighted by both studies. In addition, both found a significant association between respondents' expectations of the hospital food and their overall satisfaction with the foodservice (Fallon et al., 2008; Porter & Cant, 2009). As interventions to improve food quality were made and patient expectations were increasingly met or exceeded, patient ratings of quality increased.

Patient Population. A study by Lau and Gregoire (1998) compared perceptions of foodservice quality using questionnaire ratings between inpatients and patients who had been discharged. A 5-point scale allowed subjects to rate quality from very poor to very good. Demographic information of respondents suggested that those who completed questionnaires as inpatients were different than those who completed questionnaires after discharge. Inpatient respondents included a greater representative of older, nonwhite patients who had longer lengths of stay. In contrast, post-discharge respondents proportionately were younger, white, and had shorter lengths of stay. However, of the 48 patients who completed questionnaires both as inpatients and then post-discharge, most often no difference was found in quality ratings, indicating that it may be appropriate and effective to survey patients when they are in the hospital.

Lau and Gregoire (1998) suggest the need for future studies to focus on better ways to collect satisfaction data that are more representative of the actual patient population, as well as identifying the most efficacious and cost-effective methods. The study population was limited to one large, Midwestern hospital and no attempts were made to randomize the distribution of questionnaires.

The work of Lau and Gregoire (1998) suggest that a focus on data collection from inpatients is more convenient and cost-effective and foodservice ratings may not differ between inpatient and post-discharge patients.

2.6 Meal Rounds

Meal rounds are a tool for evaluation of inpatients food intake/ tolerance and satisfaction with foodservices. Typical meal rounds involve members of foodservice staff visiting patients during meal times and inquiring about food intake and satisfaction as well as observing tray set-up, appearance of food, and need for feeding assistance. Meal rounds can be effective in improving the quality of foodservices and can be used as a continuous quality improvement activity to readily identify nutrition risk factors (Keller, Gibbs-Ward, Randall-Simpson, Bocock, & Dimou, 2006).

A study by Watters, Sorensen, Fiala, and Wismer (2003) investigated adult patients' perceptions of hospital foodservice through inpatient meal rounds as well as focus groups with post-discharge patients and with nurses. Of 250 post-discharge patients identified and invited as potential focus group participants, only 5 patients actually attended. In contrast, for meal rounds, 135 patients were approached and 116 of these patients consented to be interviewed. For best participation, focusing on inpatients for meals rounds appears to be the most efficient (Watters et al., 2003).

Watters et al. (2003) emphasized that a team approach toward improving patient satisfaction is critical; this included having foodservice information available to all hospital staff and providing food service feedback forms on all patient units. This study also showed that meal rounds helped pinpoint the areas where foodservice staff needed additional training, such as on special or therapeutic diets, interacting with patients, offering bedside choices, and on improving communication with hospital staff. The main barrier to meal rounds is participation and support. Difficulty acquiring the support of nursing staff and other members of the disciplinary team can halt this approach (Watters et al., 2003).

2.7 Snack Room Service

Disease status, poor appetite, gastrointestinal intolerance, fatigue, early satiety and taste alterations often affect intake in a hospital setting (McLymont et al., 2003; Williams et al., 1998). Clinical dietitians often advise malnourished, hospitalized patients to consume 'small, frequent meals' or snacks every few hours to encourage oral intake and meet nutritional needs. Promoting optimal nutrition status can lead to faster recovery times and reduced hospital costs (Sullivan, Nelson, Bopp, Puskarich-May, & Walls, 1998; Giner et al., 1996; Johansen et al., 2004).

An example of a typical snack system involves snack delivery between the mid-day and evening meal and again before bedtime.

Patients may be required to make food choices well in advance of snack times using printed menus or registered dietitians may choose specific snacks based on the patient's diet order (Pietersma et al., 2003).

Common snacks include dairy products, baked goods, savory items, or fresh, canned, or dried fruit. Oral nutrition supplements (eg. Ensure or Boost) are also common hospital snack items (Price, McMurdo, & Anderson, 2006).

In a study by Pantalos and Bishop (1995), a patient-centered snack delivery system was implemented at a children's hospital. Nutritious snacks were stocked on a cart three times daily and the cart was then circulated through inpatient units. A cart attendant offered a variety of fresh and prepackaged items to patients on an increased calorie and protein diet. After implementation of the snack delivery system, the percentage of patients who actually consume the snacks they receive increased from 50% to 84%. Cost savings of more than $8,000 yearly resulted from reduced waste and foodservice labor (Pantalos & Bishop, 1995).

In a similar study, White, Wilcox, Watson, Rogany, and Meehan (2008) also evaluated a new patient-centered snack delivery system at a children's hospital. A snack cart with both savory and sweet items was operated between 2 PM and 3 PM. Patient satisfaction was determined via surveys. Surveys results showed 75% of respondents believed the cart improved foodservice satisfaction. Snack popularity was also analyzed, however, the study did not measure the effect on increased nutritional intake. Foodservice cost comparisons were made before and after implementation of the snack cart and a cost saving of 33.3% was observed (White et al., 2008).

The literature (Pantalos & Bishop, 1995; White et al., 2008; Pietersma et al., 2003) suggests that patients overwhelmingly prefer to choose food items at time of service. A patient-centered snack room service can enhance patient satisfaction and reduce costs through reductions in food waste, foodservice labor, and preventing the need for enteral or parenteral nutrition due to poor oral intake.

2.8 Room Service

Going beyond snack room service is the implementation of a full room service system, which is one of the fastest growing trends in hospital foodservice (McLymont et al., 2003). In 2008, figures from the National Society for Healthcare Foodservice Management (HFM) showed that 37 percent of respondents to an HFM survey had implemented room service dining in some way, a quarter of them within the past year (Buzalka, 2008). Many more indicated they had plans to do so in the near future.

Room service has several advantages, some of the proven top advantages include: improved patient control over food choices, improved patient satisfaction, improved food temperatures, increased foodservice employee pride in their job, decreased plate waste and decreased food cost (Buzalka, 2008; Norton, 2008; Sheehan-Smith, 2006). The main disadvantage of room service is increased cost. Increased cost mainly comes from initial investment in new equipment and computer software and initial staff training costs (Buzalka, 2008; Norton, 2008; Sheehan-Smith, 2006). A larger number of labor hours needed to run a room service system is sometimes reported, however some operations report no change in labor cost after the initial implementation period (Bulzaka, 2008). At the same time room service can reduce food cost by eliminating late trays, reducing floor stock and inventory, eliminating between meal nourishments and over production (Norton, 2008).

Deploying a Room Service System. Room service systems can vary greatly between hospitals. Norton (2008) explains that some hospitals wholly convert from a traditional trayline system to a room service system, implementing it on all compatible units and offering it to every patient capable of receiving it. Others offer it only to some areas or units, while the rest of the patients receive the traditional meal service. Patients that may not be eligible for room service are those on liquid diets, extremely restricted diets, dysphasia diets, chemically dependent and rehabilitation patients, and those who have vision, language or other limitations (Norton, 2008).

Not every hospital can afford to implement a comprehensive program turnaround; some find a compromise that provides some of the benefits of room service without all the costs. In a recent foodservice magazine article, Buzalka (2008) illustrates one example in the use of temperature-controlled mobile carts that circulate inpatient floors during lunch and dinner hours. The carts carry hot and cold meal components, which are assembled just outside patient rooms. The advantage is that patients who preorder meals in the traditional way can change their orders at mealtime. It also accommodates dietary changes and recent admissions quickly. When deploying a room service system, it is important for hospital foodservice departments to customize the system to its specific needs and to continue adapting after implementation (Buzalka, 2008).

In a qualitative study by Sheehan-Smith (2006), the best practices of hotel-style room service in hospitals were examined in four heterogeneous hospitals in the United States. The common features of hotel-style room service are meal delivery within 30 to 45 minutes, a restaurant-style menu, procedures to feed ineligible patients, tray assembly on demand, standardized employee scripting, and wait staff uniforms for room-service employees. The study found that the initial best practices included taking a multidisciplinary team approach and allowing enough time for developing and implementing the process. In order for the system to succeed, it was emphasized that employees who take patients' orders over the telephone and deliver meals should receive intense customer-service training. It was also recommended that practitioners develop a customer-driven menu and use carts with airpots for dispensing hot beverages (SheehanSmith, 2006).

The most substantial factor driving the growth of room service is competition (Buzalka, 2008; Norton, 2008). As more institutions implement the system, it becomes more difficult for others not to. This is partly because the rise in satisfaction scores at one institution negatively impacts the percentiles for other institutions (Buzalka, 2008). Patients begin to expect more and it becomes more difficult to exceed these expectations with a conventional hospital foodservice system.

In summary, the goal of this study is to use a validated, comprehensive inpatient survey to assess current foodservice satisfaction and identify the dimensions of foodservice in need of improvement. Cost-effective, customer-oriented interventions will be determined and implemented and the effect of these interventions will be monitored by re-issuing surveys. Based on survey results, it will be determined what, if any, further changes are necessary for STMH foodservice to improve satisfaction.

PART II

3. INTRODUCTION

Hospital food and nutrition services play an important role in patient recovery and well-being. Foodservice quality is also a large influence in a patient's overall hospital stay satisfaction (McLymont et al. 2003; Williams, et al., 1998). Competition in the healthcare industry is pushing administrators to innovate, become more customeroriented and improve perceived patient foodservice quality (Goehring, 2002; Kizer, 2001). One of the fastest growing trends in hospital foodservice is implementation of a room service model (Norton, 2008). Today, many conventional foodservice administrators are questioning if changing to a room service model is right for their operation and worth the cost.

In 2009, the foodservice director at St. Mark's Hospital (STMH) a 276-bed teaching hospital primarily serving the urban community of the Salt Lake Valley in Salt Lake City, Utah was faced with this situation; with pressure to increase patient satisfaction within a limited budget and the possibility of room service implementation in the very distant future.

STMH is part of the Hospital Corporation of American (HCA) and is a highly respected Mountain Star facility with patient satisfaction is its top priority. HCA is a Nashville-based hospital company with 164 hospitals in 20 states (Hospital Corporation of America, 2011). HCA conducts a quarterly, company-wide Gallup poll which surveys post-discharge patients on a variety of factors affecting their hospital stay experience and satisfaction. Gallup is a global organization that provides statistical research services for benchmarking (Gallup Organization, 2011). For the past several years, STMH, which uses an onsite centralized kitchen with a round trayline, hot thermal retention trays, cart delivery, and a selective 7-day menu cycle, had consistently scored below the company average in satisfaction in overall quality of foodservice. The single question patients are asked during the telephone interview is, "Were you satisfied with the overall quality of foodservice?" The survey poll did not provide any further details as to why many postdischarge patients were dissatisfied.

The STMH foodservice department included an internal questionnaire on the back of the patient menu. It consisted of six "yes or no" questions, with little room to add comments. The questions addressed meal taste, temperature, appearance, reliability, service, and overall satisfaction. According to department records, the questionnaire had a very low participation rate, with less than 3% of patients returning a completed survey. For those that were completed, the data was not analyzed or recorded in a standardized way. Clearly, the current questionnaire was not an adequate source of information in identifying areas of needed improvement.

In order to control costs while also improving the quality of foodservice, it is essential to isolate dimensions of foodservice that are in most urgent need of improved satisfaction and to identify precisely which effort will reap the greatest benefits (Capra et al., 2005; Dubé et al., 1994; Lau & Gregoire, 1998; Watters et al., 2006). This is also an important first step in determining whether a room service model is necessary to improve patient satisfaction and if it is worth the investment for a hospital foodservice operation (Norton, 2008; Williams et al. 1998).

Therefore, the objectives of this study were to: 1) identify the food, service, and patient variables that contribute to foodservice dissatisfaction utilizing a validated patient survey; 2) plan and implement cost-effective changes in the foodservice based on survey results; 3) monitor foodservice satisfaction scores; 4) determine and recommend further changes on to improve patient foodservice satisfaction to administrators.

4. METHODS

All methods used in this study received prior approval from St. Mark's Hospital Institutional Review Board and Utah State University Institutional Review Board.

4.1 Survey Instrument

This study used a validated survey titled The Acute Care Hospital Foodservice Patient Satisfaction questionnaire (ACHFPSQ) (Capra, Wright, & Sardie, 2005). The ACHFPSQ is based on a six-year study that sampled 2347 acute care hospital patients from two public hospitals and one private hospital. It is an accurate, reliable measure of patient foodservice satisfaction and allows quality improvement processes to be focused and measured. It differentiates foodservice into four main dimensions or categories, including: food quality, meal service quality, staff/service issues, and physical environment. Other sub-categories include: meal size, hot foods, and hunger and food quantity (Appendix).

The survey is two-pages long with 22 questions and requires 10-15 minutes to complete. Minor changes were made to the wording of some questions in order to make the survey more specific to the local area vernacular. All surveys were completed voluntarily and responses were kept completely anonymous; names were not recorded on the survey. The informed consent of patients was indicated by the return of the completed satisfaction surveys. If patients did not wish to participate in the study, a blank survey was returned.

4.2 Population and Sample

A convenience sample of 198 participants was used in two series of surveys, 99 in each of the two series of surveys. The target population consisted of inpatients served by the foodservice department during a minimum two-day stay at STMH. Both phases of survey distribution lasted 4 weeks with a goal of collecting approximately 100 completed surveys. Subjects excluded from the study included patients younger than 18 years of age and those with notable physical, cognitive, or emotional limitations as determined by the clinical dietitians. Those with an NPO diet order or those receiving enteral or parenteral nutrition as the primary source of nutrition were also excluded.

4.3 Data Collection

Clinical dietitians were responsible for the distribution and collection of surveys on a same-day basis. Surveys were distributed in-person, throughout the dietitians' shift. Patients who were visually or physically impaired and could not complete the survey independently were offered assistance. In cases of assistance, dietitians followed a structured interview protocol to reduce interviewer bias. In the first phase of the study, 166 surveys were distributed over 4 weeks and 99 completed surveys collected. The participation rate for the first phase was 60%.

4.4 Data Analysis and Interventions

Raw survey data from phase one were recorded and analyzed using the accompanying ACHFPSQ Excel spreadsheet (known as the 'ACHFPSQ automated analysis spreadsheet') (Capra et al, 2005). Question responses were coded on a scale of 1-5, representing the Likert Scale on the survey (Always=5, Often=4, Sometimes=3, Rarelly=2, Never=1). The ACHFPSQ spreadsheet calculated response means and differentiated the foodservice into the main categories and sub-categories of foodservice. The spreadsheet also allowed foodservice satisfaction to be examined further by age, selfreported diet order, gender, length-of-stay and food intake (Table 1). This initial data analysis identified the lowest-scoring foodservice categories from survey phase one and later allowed for preliminary comparison between the two survey phases, pre- and postintervention.

After initial data analysis, a committee was formed that included the Foodservice Director, head chefs, kitchen managers, and several dietitians. Based on phase one survey results, the committee decided to focus on the food quality, meal size, hunger and food quantity categories for intervention; specifically, within the standard, diabetic, and cardiac diets orders as these were the areas least positively rated.

Meal size, hunger and food quantity categories are comprised of survey questions 12, 20, 21, and 22 (Appendix). These questions address patient hunger between meals and after meals as well as patient ability to choose different sized meals and receive enough food. The food quality category is comprised of seven questions: 1, 5, 8, 9, 13, 16, and 18 (Appendix). These questions address patient food quality expectations; meal taste, variety, flavor, and texture; vegetable preparation; and the ability to choose a healthy meal.

Due to current budget limitations, interventions were restricted to low-cost improvements using current resources. All interventions were ultimately decided on by the Foodservice Director based on what he thought would be the most influential and cost-effective. It was believed that several small changes would add up to a large impact on food quality improvement. Planned interventions were made between the months of March and June 2010.

Once all interventions were running smoothly and consistently, the second phase of survey distribution was initiated. In the second phase, clinical dietitians distributed a total of 140 surveys and 99 completed surveys were collected over a four-week period. The participation rate for the second phase was 71%. Again, Excel was used to record and initially analyze raw data. It should be noted that the HCA quarterly Gallup poll results were monitored by the head foodservice director during this time.

4.5 Statistical Analysis

The raw survey data was statistically analyzed using grouped T-tests to compare the two survey phases, question by question. This was done using Excel spreadsheets imported into StatPlus for Mac LE (2009, AnalystSoft) and a group T-test was calculated for each of the 22 survey questions. For the statistical tests the significance level was set at P,0.05.

5. RESULTS AND DISCUSSION

5.1 Characteristics of Respondents

Characteristics of survey respondents for phase one and phase two are shown in Table 1. Data showed that 64% of respondents were women (n=127) and male respondents made up about one-quarter of the group (n=53). The greater number of women respondents can be partly explained by the inclusion of the maternity and women's care units in the survey. Nearly one-third of respondents were between the ages of 51-70 years (n=58). The second largest age group was 31-50 years of age, one-fifth of survey respondents (n=39). The majority of respondents (72.5%) had a length of stay less than one week at the time of the survey. Fifty-three percent of respondents reported food intake while in the hospital as 'unchanged' or 'more than usual'. The majority of respondents reported receiving a standard or diabetic diet (67.5%) and just over one-tenth reported receiving a cardiac diet (11.5%).

5.2 Assessment of Current Foodservice Satisfaction

During the first phase of survey distribution, 99 completed surveys were collected. Data analysis revealed scores for each foodservice category were very high, with Staff/Service Issues being the most positively rated (score 4.65). The least positively rated categories included Food Quality (score 4.03), Meal Size (score 4.13), Hunger and Food Quantity (score 4.21), and Hot Foods (score 4.21). Overall foodservice satisfaction was scored at 4.20 out of 5 (Table 2). Comparing satisfaction by diet order, the least positively rated diet orders based on the Food Quality category were the Standard Diet/Diabetic Diet (3.94) and Cardiac Diet (4.06) (Table 3).

Table 1. Characteristics of Respondents in Phase 1 and Phase 2

Characteristic	Survey Phase 1 #	Survey Phase 2 #	Total for Both Phases	Percent Total for Both Phases
Gender	n	n	n	%
Men	28	25	53	27%
Women	60	67	127	64%
Unreported*	11	7	18	9%
Age				
18 yrs and under	0	1	1	0.5%
19-30 yrs	16	22	38	19%
31-50 yrs	25	14	39	20%
51-70 yrs	31	27	58	29.5%
>70 yrs	14	24	38	19%
Unreported*	13	11	24	12%
Length of hospital stay to survey date				
Less than 1 week	69	74	143	72.5%
1-2 weeks	16	13	29	14.5%
More than 2 weeks	3	5	8	4%
Unreported*	11	7	18	9%
Food Intake				
Unchanged	35	42	77	39%
More than usual	10	18	28	14%
Less than usual	35	31	66	33.5%
Unreported*	19	8	27	13.5%
Self-reported diet order				
Standard/Diabetic menu	59	74	133	67.5%
Cardiac Diet	11	12	23	11.5%
Restricted for medical reasons	7	3	10	5%
Minced/Pureed/Liquid diet	2	3	5	2.5%
Don't Know	3	1	4	2%
Unreported*	17	6	23	11.5%

The ACHFPSQ defines a score of 4 out of 5 as 'very good'. These high patient satisfaction ratings for hospital foodservice may seem inconsistent with the reported perception that the quality of hospital food at STMH is below average within HCA hospitals. Capra et al. (2005) stated that the food does not have to be of a high quality for the patient to be satisfied, as satisfaction is a comparison between an expectation and a reality or experience. Patients may expect the food to be very poor, and as a result will be inclined to rate 'ordinary' food well (Capra et al., 2005). It has also been suggested that patients tend to overinflate their satisfaction depending on any number of methodological influences (Sitzia, 1999). The role of dietetic staff in distributing surveys and assisting patient in their completion where needed may have influenced patients' responses. Conversely, respondents to the company-wide Gallup poll were surveyed anonymously over the phone.

In general, the measurement and quantification of foodservice satisfaction is difficult, as it is highly subjective and influenced by many variables. In its original study, the ACHFPSQ survey instrument was only able to explain 61.2% of the variance in overall satisfaction, indicating that other variables such as pain, loss of personal power, time of year, etc., may have an effect (Capra et al., 2005). In acute care, this is not unexpected given the nature of the experience for patients.

Nevertheless, these results help establish a baseline for improvement. A score of 4 out 5 may be defined as 'very good' by the ACHFPSQ but it can be postulated that within HCA a 'very good' score is below average when compared with the other 164 HCA hospitals. These results uncover factors related to the satisfaction of STMH patients providing valuable information the hospital has never had before. One of the strengths of the ACHFPSQ is its ability to highlight very specific areas of the foodservice for quality improvement interventions such as the lowest scoring categories of foodservice (Capra et al., 2005; Fallon, Gurr, Hannan-Jones, & Bauer, 2008).

Table 2. Foodservice Satisfaction Results by Survey Category

Survey Category	Survey Phase 1 Mean*	Survey Phase 2 Mean*
Food Quality (Q1,5,8,9,13,16,18)	4.03	4.00
Meal Service Quality (Q7,10,14)	4.36	4.40
Staff/Service Issues (Q3,11,15)	4.65	4.69
Physical Environment (Q2,4,6)	4.61	4.53
Meal Size (Q12)	4.13	4.22
Hot Foods (Q17)	4.21	4.23
Hunger & Food Quantity (Q10,21,22)	4.21	4.21
Overall	4.20	4.21

*Total displayed as average score, based on Likert scale: from 1 (lowest satisfaction) to 5 (highest satisfaction)

Table 3. Average Food Quality Scores by Diet Order

Diet Order	Survey Phase 1 (n)	Survey Phase 1 Mean*	Survey Phase 2 (n)	Survey Phase 2 Mean*
Standard/ Diabetic Diet	59	3.94	74	4.07
Cardiac Diet	11	4.06	12	3.57
Restricted Diet for medical reasons (ie. Renal diet, gluten free, etc.)	7	4.34	3	4.07
Minced/Pureed/Liquid Diet	2	4.29	3	3.86
Don't Know	3	4.24	1	4.00
Unreported Diet Information**	17	-	6	-

*Total displayed as average score, based on Likert scale: from 1 (lowest satisfaction) to 5 (highest satisfaction)

**Unmarked diet order on survey

5.3 Identifying Areas of Improvement

As stated previously, improvement interventions were decided in committee. Food quantity and quality issues were discussed and many ideas were proposed for improvement. However, the STMH Foodservice Director made final decisions on specific interventions. Instead of choosing complex interventions that would significantly alter current foodservice activities, such as redesigning the snack distribution system or implementing meal rounds, the director chose to make several small changes. These small changes were quick and easy to implement with minimal influence on current foodservice activities. The goal was that several small changes would add up to a large impact on foodservice quality improvement. However, more time-intensive interventions may have had a larger impact on survey results.

5.4 Interventions

To address meal size, hunger, and food quantity issues, patients were given clearer options on the menu to control meal size. These options allowed patients to circle portion size preference: small, regular, and large. This intervention was added to the standard diet menu and the least restrictive texture-altered menu (Table 4). The foodservice staff was trained on how to modify portion sizes to meet patient requests.

The perception of food quality depends on several different attributes (Wright et al., 2006; Capra et al., 2005) and interventions were more numerous and varied, in order to address this wide range of influences represented in the seven different survey questions. To address differences among patients' unique taste and flavor preferences, the committee decided to add Mrs. Dash and hot sauce to the standard, diabetic, and cardiac diet menus as condiment options. Brown sugar and raisins were also added to the standard and cardiac breakfast menu as condiments for oatmeal.

To improve entrée food quality, several meals were evaluated by foodservice staff to identify specific issues. Problems with vegetable firmness were found and cook time was adjusted to prevent under- or overcooking of carrots and green beans. Entrée improvement examples included changing the type of noodles in the beef stroganoff entrée and refining and standardizing the seasoning in a variety of soups (Table 4).

Table 4. Low-cost Interventions Based on Survey Category

Survey Category (See Appendix)	Intervention
Meal Size, Hunger and Food Quantity (Q12,20,21,22)	• Clearer options added to menu to choose portion size • Patients circle choice: small, regular, large • Options added to standard and least restrictive texture modified diets
Food Quality (Q1,5,8,9,13,16,18)	• Decreased cook time on string beans to reduce overcooking • Increased cook time on baby carrots to reduce undercooking • Mrs. Dash and hot sauce packets added to condiment options on standard, diabetic, and cardiac diet menus • Brown sugar and raisins added as condiments to standard and cardiac breakfast menu for oatmeal • Type of noodles changed in Beef Stroganoff recipe to prevent overcooking • Taste tests and seasoning changes to a variety of soups

5.5 Monitoring Change

During the second phase of survey distribution, 99 completed surveys were collected. Initial data analysis indicated very little improvement in the categories chosen for intervention. Meal size satisfaction increased from 4.14 to 4.22 (Table 2). The hunger and food quantity category scores stayed the same after intervention at 4.21. The average score for food quality decreased slightly from 4.03

to 4.00. Overall satisfaction slightly increased post-intervention from 4.20 to 4.21. Looking specifically at food quality by diet order, the standard/diabetic diet score increased from 3.94 to 4.07 and the cardiac diet score decreased from 4.06 to 3.57. After statistical analysis of all survey questions, none of these results were shown to be statistically significant (see Table 5).

The HCA quarterly Gallup poll results were monitored by the head foodservice director during this time. The director reported no significant improvement in the foodservice satisfaction score after interventions were made.

Despite the lack of improvement in satisfaction scores, the benefits of using the ACHFPSQ as a total quality management tool is still valid (Fallon et al., 2008; Porter & Cant, 2009). STMH should continue administering the survey, at least on an annual basis, to detect any change in patient perceptions of foodservice quality. The survey tool is a vast improvement over the previous internal questionnaire and it is practical to administer within existing department resources with minimal disruption to work practices.

5.6 Future Interventions

Stiff competition in the health care industry is forcing administrators to place greater emphasis on customer-oriented service and cost-effective quality improvement (Goehring, 2002; Kizer, 2001). An important principle of customer-oriented service is creating an environment that meets or exceeds patient expectations (Fottler, Ford, Roberts, & Ford, 2000). Following strategies to improve customer-oriented service have been shown to ultimately enhance patient satisfaction (Goehring, 2002; Norton, 2008). One trend towards customer-oriented service included the deployment of a hospital room service model.

Table 5. Comparing Survey Results between Phase 1 and Phase 2

Survey Question	Phase 1[a] (n[b]=99)	Phase 2[a] (n[b]=99)	Significance[c]
1. The hospital food has been as good as I expected	3.99 ± 1.026	4.00 ± 0.968	0.945
2. The dishes and utensils are chipped and/or stained	1.31 ± 0.765	1.39 ± 0.915	0.508
3. The staff who deliver my meals are neat and clean	4.71 ± 0.704	4.81 ± 0.547	0.283
4. The hospital smells stop me from enjoying my meals	1.58 ± 0.930	1.70 ± 0.965	0.333
5. I am able to choose a healthy meal in the hospital	4.51 ± 0.800	4.40 ± 0.766	0.368
6. I am disturbed by the noise of finished meal trays being removed	1.29 ± 0.648	1.27 ± 0.607	0.812
7. The cold drinks are just the right temperature	4.24 ± 1.044	4.37 ± 0.748	0.333
8. I like the way the vegetables are cooked	3.88 ± 1.137	3.75 ± 1.023	0.401
9. The meals taste good	4.01 ± 0.941	4.04 ± 0.865	0.811
10. The hot drinks are just the right temperature	4.38 ± 0.759	4.24 ± 0.864	0.221
11. The staff who take away my finished meal tray are friendly and polite	4.72 ± 0.577	4.74 ± 0.672	0.822
12. I like to be able to choose different sized meals	4.18 ± 1.132	4.26 ± 0.966	0.597
13. The menu has enough variety for me to choose meals that I want to eat	4.23 ± 0.957	4.02 ± 0.995	0.130
14. The cold foods are the right temperature	4.53 ± 0.706	4.46 ± 0.762	0.509
15. The staff who deliver my menus are helpful	4.51 ± 0.740	4.49 ± 0.883	0.853
16. The meals have excellent and distinct flavors	3.73 ± 1.114	3.78 ± 1.013	0.724
17. The hot foods are just the right temperature	4.20 ± 0.925	4.20 ± 0.776	1.000
18. The meat is tough and dry	2.14 ± 1.055	2.09 ± 0.944	0.708
19. Overall, how would you rate your satisfaction with the foodservice	4.20 ± 0.896	4.21 ± 0.815	0.932
20. I receive enough food	4.58 ± 0.817	4.65 ± 1.017	0.577
21. I still feel hungry after my meal	1.71 ± 0.944	1.74 ± 0.750	0.859
22. I feel hungry in between meals	2.37 ± 1.141	2.33 ± 1.076	0.845

[a] A Likert-type scale was used as follows: 5=Always, 4=Often, 3=Sometimes, 2=Rarely, 1=Never (Question 19 is an exception: 5=Very good, 4=Good, 3=Okay, 2=Poor, 1=Very Poor)
[b] The actual number of participants per question varied due to missing or invalid data
[c] Results of comparison of means (T-test), $p<0.05$

The Benefits of Room Service

Hospital room service systems have several advantages, including: improved patient satisfaction, improved patient control over food choices, improved food temperatures, increased foodservice employee pride in their job, decreased plate waste and decreased food cost (Buzalka, 2008; Norton, 2008; Sheehan-Smith, 2006). The main disadvantage of room service is increased cost. Increased cost mainly comes from initial investment in new equipment and computer software, larger number of labor hours needed to run the system, and initial staff training costs (Buzalka, 2008; Norton, 2008; Sheehan-Smith, 2006). At the same time room service can reduce food cost by eliminating late trays, reducing floor stock and inventory, eliminating between meal nourishments and over production (Norton, 2008).

Hospital Competition

The most substantial factor driving the growth of room service is competition (Buzalka, 2008; Norton, 2008). As more institutions implement the system, it becomes more difficult for others not to. This is partly because the rise in satisfaction score at one institution negatively impacts the percentiles down for other institutions (Buzalka, 2008). This may be the case for St. Mark's Hospital, since several competing hospitals in the region have initiated a form of room service within the last 5 years (Intermountain Healthcare, 2011). Residents and potential patients in the Salt Lake Valley region of Utah may begin to expect more from hospital foodservice, making it more difficult to exceed these expectations with a conventional hospital foodservice system. As patient expectations increase, it becomes more difficult to meet or exceed them with a conventional hospital foodservice system.

Strategies for Deploying a Room Service System.

Room service systems can vary greatly between hospitals. From a trade journal article, Norton (2008) explains that some hospitals wholly convert from a traditional trayline system to a room service system, implementing it on all compatible units and offering it to every patient capable of receiving it. Others offer it only to some areas or units, with the rest of the patients receiving the traditional meal service. Patients that may not be eligible for room service are those on liquid diets, extremely restricted diets, dysphasia diets, chemically dependent and rehab patients, and those who have vision, language or other limitations (Norton, 2008).

In a qualitative study by Sheehan-Smith (2006), the best practices of hotel-style room service in hospitals were examined in four heterogeneous hospitals in the United States. The common features of hotel-style room service are meal delivery within 30 to 45 minutes, a restaurant-style menu, procedures to feed ineligible patients, tray assembly on demand, standardized employee scripting, and wait staff uniforms for room-service employees. The study found that the initial best practices included taking a multidisciplinary team approach and allowing enough time for developing and implementing the process. In order for the system to succeed, it was emphasized that employees who take patients' orders over the telephone and deliver meals should receive intense customer-service training. It was also recommended that practitioners develop a customer-driven menu and use carts with airpots for dispensing hot beverages (SheehanSmith, 2006).

Following best practices and observing other room service operations are helpful when deploying a hospital room service system, but it is also highly important for hospital foodservice departments to customize the system to its specific needs and to continue adapting the system after implementation (Bulzaka, 2008).

6. CONCLUSIONS AND APPLICATIONS

Based on the follow-up survey results and recent quarterly Gallup poll, low-cost interventions were not shown to be effective in significantly increasing foodservice satisfaction at St. Mark's Hospital. Findings may indicate that the STMH foodservice department is reaching its patient-satisfaction potential within its current conventional system and cost-limitations.

Patient satisfaction is the highest priority at STMH and foodservice quality plays a large roll in the overall hospital stay experience (McLymont et al., 2003; Williams et al., 1998). For considerable improvement in foodservice satisfaction, it may be necessary for STMH foodservice administrators to consider higher-cost interventions. Based on this study, it was recommended that a room service system be considered as a way to truly meet and possibly exceed the foodservice expectations of patients.

Traditional hospital foodservice operations with limited funds that are considering options to increase patient foodservice satisfaction may consider the following recommendations:

1) Utilize a validated patient survey, such as The Acute Care Hospital Foodservice Patient Satisfaction questionnaire (ACHFPSQ), to assess current patient foodservice satisfaction.

2) Focus improvement interventions on the lowest scoring foodservice categories.

3) Improvement interventions may be decided in committee, keeping in mind the varied dimensions of foodservice and examples of successful interventions at other institutions.

4) Monitor change to patient foodservice satisfaction score by re-issuing surveys.

5) Make continuous quality improvements; survey and intervention periods may be repeated indefinitely.

6) If significant improvement to patient foodservice satisfaction is not made, survey findings may be helpful in validating and securing additional funding for highercost interventions, such as a version of a room service system.

7) When higher funding or deployment of a room service system is not an option for the operation, it is recommended to remain focused on continuous quality improvements over time with revised and realistic goals of improving patient foodservice satisfaction given the operations limitations.

7. LIMITATIONS AND FUTURE RESEARCH

Study participants were limited to a convenience sample of 198 patients in one hospital. It is likely that patients, who are most severely ill, have limited literacy skills and those from non-English speaking backgrounds are underrepresented.

Patients were only surveyed one time during their hospital stay, typically within a couple days of admission and it is possible that patients' opinions of foodservice satisfaction changed over the course of their stay. Several studies report longer length of stay is associated with less satisfaction with foodservice (Lau & Gregoire, 1998; Stanga et al., 2003). However, a study by Fallon, Gurr, Hannan-Jones, & Bauer (2008) found no statistically significant difference between overall satisfaction and length of stay. The hospital in the study offers long stay patients (≥ 14 days) an alternate menu with an additional eight choices for both lunch and evening meals. Fallon et al. (2008) asserts this may explain why there was no significant decrease in satisfaction with increased length of stay.

Response rates ranged from 60-71%, however, this is consistent with response rates from foodservice satisfaction surveys in the literature (Wright et al., 2006). One intervention period was implemented during this study and interventions were not exhaustive or standardized over many intervention periods. If this study were repeated it would be preferable that several intervention and survey periods be implemented over a longer time period. The interventions chosen were quick and easy to add within the current foodservice system, a longer time period would allow for more complex interventions to be designed, such as snack room service or patient meal rounds. These more time-intensive interventions may have had a larger impact on survey results.

Future research might focus on investigating whether most patients now expect hotel-style room service during hospital stays and what has influenced their expectations. Further exploration into what influences patient expectations in a given region may be helpful in determining whether a hospital must deploy a room service system in order to compete and meet patient expectations.

PART III

8. CONCLUSIONS AND APPLICATIONS

Patient satisfaction is the highest priority at St. Mark's Hospital (STMH) and foodservice quality plays an important role in the overall hospital stay experience (McLymont et al., 2003; Williams et al., 1998). However, in the quarterly company-wide poll, STMH consistently scored below the company average in satisfaction in overall quality of foodservice. The purpose of this study was to use a validated patient survey to: 1) identify the food, service, and patient variables that contribute to foodservice dissatisfaction; 2) plan and implement cost-effective changes in the foodservice based on survey results; 3) monitor foodservice satisfaction scores; and 4) determine and recommend further changes on to improve patient foodservice satisfaction to administrators.

The study was conducted at STMH, a 276-bed acute care hospital in Salt Lake City, Utah, over a six-month period from February 2010 to July 2010. The STMH foodservice department operates as a conventional centralized kitchen. This traditional system uses a round trayline, hot thermal retention trays, cart delivery, and a selective 7-day menu cycle. The Acute Care Hospital Foodservice

Patient Satisfaction Questionnaire (ACHFPSQ) was chosen as the validated survey instrument because of its accuracy and reliability in measuring patient foodservice satisfaction (Capra et al., 2005).

Clinical dietitians were responsible for the distribution and collection of surveys on a same-day, convenience-basis. Ninety-nine completed surveys were collected in the first phase of distribution. Data analysis of the first survey phase revealed high satisfaction scores in all foodservice categories and in overall satisfaction with foodservice. Nevertheless, the results were helpful in establishing a baseline for improvement and for identifying the lowest scoring foodservice categories. The least positively rated categories were food quality, meal size, hunger and food quantity. Lowcost interventions were decided in committee and implemented to address these categories. The second phase of surveys was distributed to monitor the effect of these low-cost interventions on patient foodservice satisfaction. An additional 99 surveys were collected in the second phase for a total of 198 survey participants. Grouped T-tests were calculated to compare the two survey phases and it was found that there was no statistically significant improvement in patient foodservice satisfaction. The company Gallup poll was also monitored during this time and no significant improvement in patient satisfaction with foodservice was found.

Despite the lack of statistically significant change in satisfaction scores, the benefits of using the ACHFPSQ as a total quality management tool is still recommended (Fallon et al., 2008; Porter & Cant, 2009). STMH should continue administering the survey, at least on an annual basis, to detect any change in patient perceptions of foodservice quality. A survey period could be held in February and results could then be reported and addressed in March as part of the American Dietetic Association National Nutrition Month. The survey tool is a vast improvement over the previous internal questionnaire and it is practical to administer within existing department resources with minimal disruption to work practices. As hospitals in the private sector compete for patients' care, Fallon et al. (2008) suggests results from the survey may be used as a marketing tool to differentiate a hospital on the basis of the quality of the foodservice. Results can also be used to provide positive feedback to foodservice workers as a way to contribute to job satisfaction (Fallon et al., 2008).

Based on the follow-up survey results and most recent quarterly company poll, the low-cost interventions selected for this study were not shown to be effective in significantly increasing foodservice satisfaction at STMH. Findings may suggest that the STMH foodservice department is reaching its patient-satisfaction potential within its current conventional system and cost-limitations. For considerable improvement in foodservice satisfaction, it may be necessary for STMH administrators to consider higher-cost, customer-oriented investments, such as a room service system.

The link between a customer-oriented service culture and patient satisfaction is becoming clear. One of the most important principles in customer-oriented service is creating an environment that meets or exceeds patient expectations (Fottler, Ford, Roberts, & Ford, 2000). Room service systems are one of the fastest growing trends in hospital foodservice due in part to the ability it gives foodservice institutions to exceed patient expectations (McLymont et al., 2003). In 2008, figures from the National Society for Healthcare Foodservice Management (HFM) showed that 37 percent of respondents to an HFM survey had implemented room service dining in some way, a quarter of them within the past year (Buzalka, 2008). Many more indicated they had plans to do so in the near future.

Hospital room service systems have several large advantages, including: drastically improved patient satisfaction, improved patient control over food choices, improved food temperatures, increased foodservice employee pride in their job, decreased plate waste and decreased food cost (Buzalka, 2008; Norton, 2008; Sheehan Smith, 2006; McLymont et al., 2003). It has been well documented that patients prefer to choose food items at time of service (Pantalos & Bishop, 1995; White et al., 2008; Pietersma et al., 2003).

The main disadvantage of room service is increased cost. Increased cost mainly comes from initial investment in new equipment and computer software and initial staff training costs (Buzalka, 2008; Norton, 2008; Sheehan-Smith, 2006). A larger number of labor hours needed to run a room service system is sometimes reported, however some operations report no change in labor cost after the initial implementation period (Bulzaka, 2008). At the same time room service can reduce food cost by eliminating late trays, reducing floor stock and inventory, eliminating between meal nourishments and over production (Norton, 2008). STMH can confront the cost challenges by conducting a cost-benefit study. This type of study could research and weigh the initial room service cost with the overall financial and patient satisfaction benefit. A break-even point may then be determined and a cost-savings may be estimated over time.

Room service systems can vary greatly between hospitals. From a trade journal article, Norton (2008) explains that some hospitals wholly convert from a traditional trayline system to a room service system, implementing it on all compatible units and offering it to every patient capable of receiving it. Others offer it only to some areas or units, with the rest of the patients receiving the traditional meal service. Patients that may not be eligible for room service are those on liquid diets, extremely restricted diets, dysphasia diets, chemically dependent and rehab patients, and those who have vision, language or other limitations (Norton, 2008).

Hours of room service operation, including evening service hour cutoffs are an additional challenge. According to Bulzaka (2008), most programs begin serving around 6:30-7 am and go into the early evening until 6:30-7 pm. However, some programs run 24 hours a day, seven days a week. Hours of operation should be decided based on patient population; most important is that the majority of patients have what they want to eat when they want to eat it (Norton, 2008).

In a qualitative study by Sheehan-Smith (2006), the best practices of hotel-style room service in hospitals were examined. The common features of hotel-style room service are meal delivery within 30 to 45 minutes, a restaurant-style menu, procedures to feed ineligible patients, tray assembly on demand, standardized employee scripting, and wait staff uniforms for room-service employees. The study found that the initial best practices included taking a multidisciplinary team approach and allowing enough time for developing and implementing the process. In order for the system to succeed, it was emphasized that employees who take patients' orders over the telephone and deliver meals should receive intense customer-service training.

It was also recommended that practitioners develop a customer-driven menu and use carts with airpots for dispensing hot beverages (Sheehan-Smith, 2006). Following best practices and observing other room service operations are helpful when deploying a hospital room service system, but it is also highly important for hospital foodservice departments to customize the system to its specific needs and to continue adapting the system after implementation (Bulzaka, 2008).

The most substantial factor driving the growth of room service is competition (Buzalka, 2008; Norton, 2008). As more institutions implement the system, it becomes more difficult for others not to. This is partly because the rise in satisfaction scores at one institution negatively impacts the percentiles at other institutions (Buzalka, 2008). St. Mark's Hospital may be facing this situation, since several other competing hospitals in the region have initiated a form of room service within the last 5 years (Intermountain Healthcare, 2011). Residents and potential patients in the Salt Lake Valley region of Utah are conceivably beginning to expect more from hospital foodservice, making it more difficult to exceed these expectations with a conventional hospital foodservice system.

In summary, due to the highly competitive nature of the health care industry, organizations are becoming more customer-oriented and patients are beginning to expect more (Kizer, 2001). Further investments in the conventional foodservice system likely will not achieve the goal of significant improvement in patient satisfaction. Based on this study, it was recommended that a room service system be considered as a way to truly meet and possibly exceed the foodservice expectations of patients. To accomplish this practioners may consider following the best practices of hotel-style room service in hospitals as outlined by Sheehan-Smith (2006), using a multidisciplinary team approach, and observing current room service operations in the area.

In the summer of 2011, the St. Mark's Hospital foodservice department will begin implementation of a hotel-style room service system. The department was able to secure room service funding from hospital administration due in part to the findings of this study.

Traditional hospital foodservice operations with limited funds that are considering options to increase patient foodservice satisfaction may consider the following recommendations:

1) Utilize a validated patient survey, such as The Acute Care Hospital Foodservice Patient Satisfaction questionnaire (ACHFPSQ), to assess current patient foodservice satisfaction.

2) Focus improvement interventions on the lowest scoring foodservice categories.

3) Improvement interventions may be decided in committee, keeping in mind the varied dimensions of foodservice and examples of successful interventions at other institutions.

4) Monitor change to patient foodservice satisfaction score by re-issuing surveys.

5) Make continuous quality improvements; survey and intervention periods may be repeated indefinitely.

6) If significant improvement to patient foodservice satisfaction is not made, survey findings may be helpful in validating and securing additional funding for highercost interventions, such as a version of a room service system.

7) When higher funding or deployment of a room service system is not an option for the operation, it is recommended to remain focused on continuous quality improvements over time with revised and realistic goals of improving patient foodservice satisfaction given the operations limitations.

9. LIMITATIONS AND FUTURE RESEARCH

Study participants were limited to a convenience sample of 198 patients in one hospital. It is likely that patients, who are most severely ill, have limited literacy skills and those from non-English speaking backgrounds are underrepresented. Response rates ranged from 60-71%, however, this is consistent with response rates from foodservice satisfaction surveys in the literature (Wright et al., 2006).

Patients were only surveyed one time during their hospital stay, typically within a couple days of admission and it is possible that patients' opinions of foodservice satisfaction changed over the course of their stay. Several studies report longer length of stay is associated with less satisfaction with foodservice (Lau & Gregoire, 1998; Stanga et al., 2003). However, a study by Fallon, Gurr, Hannan-Jones, & Bauer (2008) found no statistically significant difference between overall satisfaction and length of stay. The hospital in the study offers long stay patients (≥ 14 days) an alternate menu with an additional eight choices for both lunch and evening meals. Fallon et al. (2008) asserts this may explain why there was no significant decrease in satisfaction with increased length of stay.

One intervention period was implemented during this study and interventions were not exhaustive or standardized over many intervention periods. If this study were repeated it would be preferable that several intervention and survey periods be implemented over a longer time period. In the study by Fallon et al. (2008), the ACHFPSQ was administered annually over a 3-year period. Improvement interventions were carefully chosen and thorough. Continuous quality improvement became a collaborative team approach within the department.

It is also important to note the suggested tendency of hospital patients overinflating their satisfaction ratings depending on any number of methodological influences (Sitzia, 1999). The role of dietetic staff in distributing surveys and assisting patients in their completion where needed may have influenced patients' responses. Conversely, respondents to the company-wide Gallup poll where surveyed anonymously over the phone.

In general, the measurement and quantification of foodservice satisfaction is difficult, as it is highly subjective and influenced by many variables. In its original study, the ACHFPSQ survey instrument was only able to explain 61.2% of the variance in overall satisfaction, indicating that other variables such as pain, loss of personal power, time of year, etc. have an effect (Capra et al., 2005). In acute care, this is not unexpected given the nature of the experience for patients.

Future research might focus on investigating whether most patients now expect hotel-style room service during hospital stays and what has influenced their expectations. Further exploration into what influences patient expectations in a given region may be helpful in determining whether a hospital must deploy a room service system in order to compete and meet patient expectations.

REFERENCES

1. Avis, M., Bond, M., & Arthur, A. (1995). Satisfying solutions? A review of some unresolved issues in the measurement of patient satisfaction. Journal of Advanced Nursing, 22, 316-322.

2. Buzalka, M. (2008, December 1). You've tossed your trayline...now what? Food Management, pp.32-35.

3. Capra, S., Wright, O., Sardie, M., Bauer, J., & Askew, D. (2005). The acute hospital foodservice patient satisfaction questionnaire: The development of a valid and reliable tool to measure patient satisfaction with acute care hospital foodservices. Foodservice Research International, 16, 1-14.

4. Drain, M. (2001). Quality improvement in primary care and the importance of patient perceptions. The Journal of Ambulatory Care Management, 24, 30-46.

5. Draper, M., Cohen, P., and Buchan, H. (2001). Seeking consumer views: what use are results of hospital patient satisfaction surveys? International Journal for Quality in Health Care, 13(6), 463-468.

6. Dubé, L., Trudeau, E., & Belanger, M. C. (1994). Determining the complexity of patient satisfaction with foodservices. Journal of the American Dietetic Association, 98, 394-399.

7. Fallon, A., Gurr, S., Hannan-Jones, M., & Bauer, J. D. (2008). Use of the acute care hospital foodservice patient satisfaction questionnaire to monitor trends in patient satisfaction with foodservice at an acute care private hospital. Nutrition & Dietetics, 65, 41-46.

8. Fottler, M. D., Ford, R. C, & Roberts, V., Ford, E. W., Spears, J. D. (2000). Creating a healing environment: the importance of the service setting in the new consumeroriented healthcare system. Journal of Healthcare Management, 45, 91-106.

9. Gallup Organization (2011). About Gallup. Retrieved March 5, 2011, from http://www.gallup.com/corporate/115/about-gallup.aspx.

10. Giner, M., Laviano, A., Meguid, M. M., & Gleason, J. R. (1996). In 1995 a correlation between malnutrition and poor outcome in critically ill patients still exists. Nutrition, 12, 23-29.

11. Goehring, K. S. (2002). Linking a service culture with patient satisfaction. Healthcare Executive, 17, 60-61.

12. Hospital Corporation of America (2011). About HCA. Retrieved March 25, 2011, from http://www.hcahealthcare.com/about/.

13. Intermountain Healthcare (2011). Search Intermountain. Retrieved March 5, 2011, from http:// intermountainhealthcare.org /Pages/ home.aspx.

14. Johansen, N., Kondrop, J., Plum, L. M., Bak, L., Norregaard, P., Bunch, E., Baernthsen, H., Andersen, J. R., Larsen, I. H., & Martinsen, A. (2004). Effect of nutritional support on clinical outcome in patients at nutritional risk. Clinical Nutrition, 23, 539-550.

15. Keller, H. H., Gibbs-Ward, A., Randall-Simpson, J., Bocock, M.A., & Dimou, E. (2006). Meal rounds: an essential aspect of quality nutrition services in long-term care. Journal of the American Medical Directors Association, 7(1), 40-45.

16. Kizer, K. W. (2001). Establishing health care performance standards in an era of consumerism. Journal of the American Medical Association. 286(10), 1213-1217

17. Lau, C. & Gregoire, M. B. (1998). Quality ratings of a hospital foodservice department by inpatients and postdischarge patients. Journal of the American Dietetic Association, 98, 1303-1307.

18. McLymont, V., Sharon, C., & Stell, F. (2003). Improving patient meal satisfaction with room service meal delivery. Journal of nursing care quality, 18(1), 27-37.

19. Norton, C. (2008). Why room service? Is it for your hospital's foodservice operation? Market Link, 27, 1-11.

20. Pantalos, D. C. & Bishop, C. B. (1995). A patient centered system for snack delivery. Journal of the American Dietetic Association, 95, A39.

21. Pietersma, P., Follet-Bick, S., Wilkinson, B., Guebert, N., Fisher, K., & Pereira, J. (2003). A bedside food cart as an alternative food service for acute and palliative oncological patients. Support Care Cancer, 11: 611-614.

22. Price, M. J. G., McMurdo, M. E. T., & Anderson, A. S. A personalized snack-based intervention for hip-fracture patients: development, feasibility, and acceptability. Journal of Human Nutrition and Dietetics, 19, 139-145.

23. Porter, J. & Cant, R. (2009). Exploring hospital patients' satisfaction with cook-chill foodservice systems: a preliminary study using a validated questionnaire. Journal of Foodservice, 20, 81-89.

24. Sheehan-Smith, L. (2006). Key facilitators and best practices of hotel-style room service in hosptials. Journal of the American Dietetic Association, 106, 581-586.

25. Sitzia, J. (1999). How valid and reliable are patient satisfaction data? An analysis of 195 studies. International Journal for Quality in Health Care, 11, 319-328.

26. Stanga, Z., Zurfluh, Y., Roselli, M., Sterchi, A. B., Tanner, B., & Knecht, G. (2003). Hospital food. a survey of patients' perceptions. Clinical Nutrition, 22, 241-246.

27. Sullivan, D. H., Nelson, C. L., Bopp, M. M., Puskarich-May, C. L., & Walls, R. C. (1998). Nightly enteral nutrition support of elderly hip fracture patients: a phase I trial. Journal of the American College of Nutrition, 17, 155-161.

28. Urden, L. D. (2002). Patient satisfaction measurement: current issues and implications. Lippincott's Case Management, 7(5), 194-200.

29. Vozenilek, G. P. (1999). How to enhance a hospital foodservice: giving patients the gourmet treatment. Journal of the American Dietetic Association, 99, 1048-1049.

30. Watters, C. A., Sorensen, J., Fiala, A,. & Wismer, W. (2003). Exploring patient satisfaction with foodservice through focus groups and meal rounds. Journal of the American Dietetic Association, 103, 1347-1349.

31. White, M., Wilcox, J., Watson, R., Rogany, A. & Meehan, L. (2008). Introduction of a patient-centered snack delivery system in a children's hospital increases patient satisfaction and decreases foodservice costs. Journal of Foodservice, 19, 194-199.

32. Williams, R., Virtue, K., & Adkins, A. (1998). Room service improves patient food intake and satisfaction with hospital food. Journal of Pediatric Oncology Nursing, 15(3),183-189.

33. Wright, O. R. L., Connelly, K. B., & Capra, S. (2006). Consumer evaluation of hospital foodservice quality: an empirical investigation. International Journal of Health Care Quality Assurance, 19(2), 181-194.

5

SERVICE QUALITY MANAGEMENT IN HOTEL INDUSTRY: A CONCEPTUAL FRAMEWORK FOR FOOD AND BEVERAGE DEPARTMENTS

Ala`a Nimer AbuKhalifeh & Ahmad Puad Mat Som

1 School of Housing, Building and Planning, Universiti Sains Malaysia, Penang, Malaysia

ABSTRACT

Service quality has been an important subject of research involving food and beverage (F&B) departments of hotels. Despite a substantial number of studies on service quality, the reasons why guests revisit a hotel and why a high-quality service from the F&B department is needed have remained unanswered. This paper aims to review existing literature on service quality management in the F&B departments of hotels, its process, and the effective service quality management framework. This paper discusses famous models, and explains Parasuraman's dimensional framework of service quality management in the area of F&B and its application to the hotel industry. The conceptual paper suggests application of the dimensional model in the F&B department and encourages hotels to improve its management to better satisfy their guests.

KEYWORDS

service quality, hotel industry, parasuraman dimensional model, food and beverage department

1. INTRODUCTION

Hotel management scholars consider service quality a precedent to guest satisfaction. Some experiential studies show that getting a good quality of service from the food and beverage (F&B) department of a hotel is important for many guests (Armstrong et al., 1997; Crick & Spencer, 2011; Cronin & Taylor, 1992a; Getty & Thompson, 1994; Lam & Zhang, 1998; Parasuraman. et al., 1988a; Sulek & Hensley, 2010; Zeithaml & Bitner, 2003). Researchers have defined service quality in relation to the concept of consumer-perceived quality, which is based solely on the perspective of customers. Such a perception is built in a place where an organization supplies goods and services to customers in a manner that satisfies them and where they examine service quality (Babajide, 2011, p. 48; Carman, 1990; Choi & Chu 1998; Christie 2002; Cronin et al., 2000; Gounaris et al., 2003; Mei et al., 1999; Mill, 2002; O'Neill, 2001; Oberoi & Hales, 1990; Presbury et al., 2005; Qu & Tsang, 1998; Rust & Zahorik, 1993; Saleh & Ryan, 1991; Zeithaml et al., 1996). In the hotel industry, service quality that satisfies customers is important to encourage them to revisit and to earn their loyalty (Carev, 2008; Carman, 1990; Jabulani, 2001; Parasuraman. et al., 1988a; Zeithaml & Bitner, 2003) and satisfaction (Babajide, 2011, p. 48; Christie 2002; Hersh, 2010, p. 209; Ladhari, 2009, p. 311; Oliver, 1999).

Parasuraman et al. (1985; 1988b), Qu and Tsang (1998), and Zeithaml et al. (1996) define "perceived service quality" as the global evaluation by customers of the overall excellence or superiority of a service. The definition is similar to the concept of attitude. Based on the exploratory definition of service quality as perceived by customers, it is the gap (discrepancy) between a customer's expectation of a service and his perceptions of the actual service received (p. 254 - 255). Much of the recent research on service quality has been carried out within the framework of the SQ (service quality) model developed from the extensive research of Parasuraman et al. (1990, 1985, and 1988b), Qu & Tsang (1998), and Zeithaml et al. (1996). Service quality has been discussed in various literatures, and one of the most largely used models measuring service quality is the SERVQUAL.

Taking on views from the study by Parasuraman et al. (1995, 1998), this study suggests that service quality and its dimensions are among the important factors influencing guest satisfaction. Considering the dimensions of service quality is, therefore, necessary as well. This dimension was focused on several studies on which hotel administrators received the support to implement the plan for improving service quality (AlRousan, 2011; Mohsin, 2011; Parasuraman et al., 1985). The benefits of a good service quality as far as the hotel industry is concerned include greater guest satisfaction and enhancement of employee performance (Cronin et al., 2000; Duncan & Elliott, 2002; Johnson et al., 1995; Kang & James, 2004; Lidén, 2003; Peter & Nicole, 2006; Rust et al., 1995; Soutar et al., 2001; Su, 2004). Zeithaml and Bitner (2003) mentioned that perception of service quality becomes global depending on the prevalence of the service. The idea that service quality is a multidimensional concept is commonly accepted.

According to Antony and Ghosh (2004) and Harvey (1998), service quality covers meeting and exceeding customer expectations. This description has become popular since its inception in the late 1970s. The concept of service quality stemmed from the area of marketing, which puts importance to the relationship between a business and its customers. Service quality is intangible, making its measurement impossible. Evaluation of service quality largely depends on management of appearances and perceptions (p. 380).

This paper reviews existing literature and suggests a theoretical framework for F&B department that illustrates the evaluation of service quality in the hotel industry in Jordan. It uses the five dimensions developed by Parasuraman et al. (1985) and Zeithaml et al. (1996). These five dimensions are the followings: (1) tangibles – refer to physical facilities, equipment, appearance of personal, etc.; (2) reliability – ability to dependably and accurately perform the promised service; (3) responsiveness – willingness to help customers and provide prompt service; (4) assurance – knowledge and courtesy of employees, and their ability to convey trust and confidence; and (5) empathy – care for and individualized attention to guests.

2. THEORIES AND MODELS IN SERVICE QUALITY

There are five dimensions of service quality, namely, tangibility (Cronin & Taylor, 1992b; Kang & James, 2004; Saleh & Ryan, 1991; Zeithaml & Bitner, 2003), reliability (AlRousan, 2011; Kotler & Bloom, 1984; Ladhari, 2009, p. 311; Parasuraman. et al., 1988a), responsiveness (Antony & Ghosh, 2004; Cheung & Law, 1998; Knuston et al., 1991; Upal, 2008), assurance (Al Khattab, 2011; Cronin & Taylor, 1992a; Pitt & Jeantrout, 1994; Powapaka, 1996), and empathy (Harvey, 1998; Mohsin, 2011; Parasuraman. et al., 1988a; Qu & Nelson, 1998; Zeithaml et al., 1993). Several studies have identified SERVQUAL, but Parasuraman et al. (1985) were the first to point out that the direction of marketing services is one of the considerations in guest administration. Service quality, such as in the hotel industry, is difficult to understand compared with product quality because the former is intangible. Several studies reported that the SERVQUAL scale is not universal because the dimensionality of service quality apparently depends on the type of service examined (Ladhari, 2009, p. 311).

There has been variable help for the validity of the SERVQUAL model in the hotel service industry (Ladhari, 2009, p. 311; Wilkins et al.,

2007). Akan (1995) explores the applicability of the SERVQUAL model in the Turkish hospitality industry and identifies seven dimensions in the model: (1) courtesy and competence of the person, (2) communication and transactions, (3) tangibles, (4) knowledge and understanding of the customer, (5) accuracy and speed of service, study of the hotel industry, (6) solutions to problems; and (7) accuracy of hotel reservations. Mei et al. (1999) determine the dimensions of service quality in the Australian hotel industry and develop a new scale of service quality in the hospitality industry, called "HOLSERV," with three dimensions: (1) employees, (2) tangibles, and (3) reliability. Saleh and Ryan (1991) report five dimensions of service quality that differ from those in the SERVQUAL model: (1) conviviality; (2) tangibles; (3) reassurance; (4) avoidance of sarcasm; and (5) empathy. Oberoi and Hales (1990) suggest that service quality plays an important role in conference hotels in the United Kingdom and report that service-quality perception has two dimensions: (1) tangibles; and (2) intangibles.

Getty and Thompson (1994) describe the validity of the SERVQUAL dimensions in the hotel sector and report that only two of the dimensions (tangibles and reliability) are generic. Getty and Getty (2003) test the dimensions of service quality in the hotel industry in the United States and develop a new scale (called "lodging quality index") with five service-quality dimensions: (1) tangibility, (2) reliability, (3) responsiveness, (4) confidence, and (5) communication. Al Roussan (2011) examines Parasuraman model for hotel service quality and customer loyalty in Jordanian hotels using the Marriott hotel chain as a case study. The study discusses the five gaps in service quality in the context of Jordan and applies this model to the Marriott hotel chain to test its quality of service. The benefit of this framework is applying the elements of "various models" into a clarified hotel and tourism industry management as Parasuraman (1995, p. 1) says:

"Attainment of quality in products and services has become a pivotal concern in 1980 while the quality in tangible goods has been described and measured by marketers, quality of service is largely undefined and unresearched. The authors attempt to rectify this situation by reporting the insight obtained in an extensive exploration of quality in four service businesses and developing a model of service quality".

3. SERVICE QUALITY DIMENSIONS

The first definition, reliability is defined as the ability of a service supplier to promptly deliver a good quality of service. Observing service delivery contracts, pricing, and complaints handling are all important for a business. (AlRousan, 2011; Parasuraman et al., 1985; Tat & Raymound, 1999). The second concept, responsiveness is the agreeableness to help guest by providing immediate service as soon as a request is received. The service guests will provide soon and prompt attention to all their guest requests, question and suggestion (Al Rousan, 2011; Knutson et al., 1990; Parasuraman et al., 1985). The third concept, assurance is with consideration to the service quality detention which considerates on the capability to activate trust and confidence of the product or service provided. The fourth dimension empathy is the service quality manner that stresses on the contacting of guest's as personalized. The end of this step, tangibility is the service dimension that makes a product or service practical and usable for customers (AlRousan, 2011; Parasuraman et al., 1985).

Figure 1 shows the Parasuraman dimensional SERVQUAL management framework.

3.1 Guest Satisfaction

Guest satisfaction is the evaluation of a guest whether or not the quality of a service meets his expectations. Performance that falls below expectations makes guests dissatisfied (Kottler & Killer,

2006). In the hotel industry, satisfaction of guests is in terms of their satisfaction with the service of the F&B department and the customer relations developed by hotel staff (Al Rousan, 2011; Parasuraman et al., 1985). Customer satisfaction is considered one of the most important outcomes of all marketing activities in a market-oriented firm. The obvious objective of satisfying a customer is to expand a business, to gain higher market share, and to acquire repeat and referral business, all of which lead to improved profitability (Barsky, 1992; Carev, 2008). However, the guest satisfaction concept in this model refers to judgement that depends on both good service quality and fitness of service to the needs of guests (Tse and Wilton, 1988). According to Parasuraman, guest satisfaction is the level in which expectation of a guest is met or exceeded. Carev (2008) defines service quality satisfaction based on overall consideration by a guest of the advantages of a service taking into account perceptions of what is reserved and what is given (Parasuraman. et al. 1988b). Service providers should be able to understand the behavior and needs of guests to be able to know how to satisfy them. Cheung and Law (1998) develop a new model for the measurement of hospitality SQ that incorporates performance of hotel employees in the SQ level. This new SQ for the customer satisfaction model, in turn, can be directly integrated into the hospitality TQ operation because SQ is a basic component of TQ. This new model prevails over the traditional SQ models in several areas. The new model is more comprehensive. Moreover, the new model better reflects the actual SQ situation that fits the TQ concept as both TQ and staff performance aim at sustained improvement. The former steers towards continual improvement of each process, whereas the latter is directed towards continual improvement in performance. (p. 402).

This study is one of the studies on the quality of the F&B department in Jordanian hotels and its relationship with guest satisfaction and service quality. This study suggests a model that may be successfully applied to service quality. Furthermore, this study suggests the application of a guest satisfaction model that hotel managers may find useful. Finally there was an absence of competent data to design and develop hotel service quality and guest satisfaction. To benefit from this study, F&B department and hotel management will develop and improve their service and guest satisfaction. Nevertheless, this study is expected to help F&B departments and management of hotels to develop and improve their service and thus better satisfy their guests.

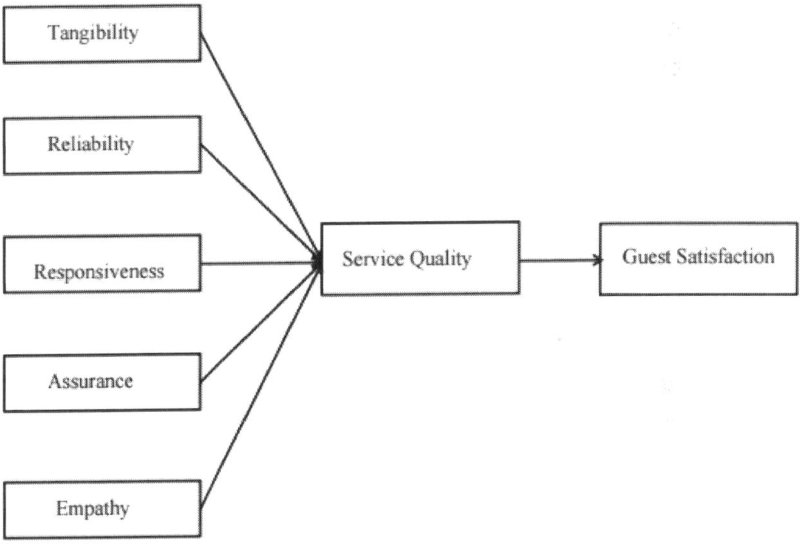

Figure 1. Parasuraman SERVQUAL management framework: adopted by Parasuraman et al. (1985)

4. SERVICE QUALITY IN FOOD AND BEVERAGE DEPARTMENTS

F&B department in the hotel industry has received considerable attention from researchers and academics given that this department may help guide the development of service quality in the entire F&B industry (Arizon, 2010, p. 122). Understanding the antecedent of service quality" allows the hotel industry to concentrate on SERVQUAL models. Reviewing more than 32 empirical studies, Arizon (2010, p. 122) focuses on service quality delivery in the F&B industry in the western Cape, evaluates factors that affect service quality, and explains SERVQUAL application and gaps in the hotel industry. Cho and Wong (1998 p. 285) show the relationship between service quality and guest satisfaction, and discuss the measurement of service quality using a computerized food and beverage cost control system. According to their study, "The gaps between management requirements and the functionality of computer systems could thereby be reduced. The average scores in terms of the expectations of all attributes are high, suggesting that there are few redundant functions/features in the existing FBCCSs in Hong Kong hotels." In another study, the Gui Hilton Helsinki Airport hotel is used as the main antecedent of breakfast service quality, which covers quality of food and reliability of service in restaurants. The gap between consumer expectation and management perception, as well as the gap between service quality specification and service delivery leads to a gap in perceived service quality (Babajide, 2011, p. 48). Ruetzler (2008) has three studies on food and beverage, namely, "Service Quality and Culture," "Service Quality and School Food Service," and "Service Quality Model."

Nield et al. (2000) affirm "the role and importance of food service in tourist satisfaction and the differences in satisfaction levels among regional groups. Analysis of the findings reveals that significant differences existed between satisfaction perception of three tourist groups on value for money, quality of food, a number of dishes, food service standards, variety of dishes, and presentation of food and speed of service in general, bread, coffee, meat, and soup in particular. Overall results further indicate that quality of food, value for money, variety of dishes, attractiveness of the surroundings, and presentation of food were the attributes that most affected the overall food service experience in Romania (p. 375). Result of their study show that food service is an important contributor to tourist satisfaction, and that there are significant differences in satisfaction levels with food service between eastern and western European, and Romanian tourists. They confirm that guest satisfaction plays a very important role in the hotel and tourism industry.

In the study by Sriyam (2010, pp. 7-17), the use of the SERVQUAL approach to measure service quality in hotel and other service industries is accepted. The study explores expectation and perception levels of customers towards service quality that focuses on the front office staff and service quality dimensions in other hotel departments, such as that of F&B. Moreover, the SERVQUAL model scale enables actual measurement of service quality. The model on which SERVQUAL is based allows customers to evaluate quality of a service based on five distinct dimensions: tangibles, reliability, responsiveness, assurance, and empathy. The provision of individualized attention to customers and the use of this model are suggested by previous studies (e.g., Lin, 2006; and Nield et al., 2000). In existing literatures, theories and models for measuring service quality in food and beverage departments are recommended and recognized as indicators of SERVQUAL (Al Khattab, 2011a; Al Rousan, 2011; Christie 2002; Gazzoli et al., 2009; Getty & Getty, 2003; Gronroos, 1990; Knutson et al., 1990; Mill, 2002; Oliver, 1980; Qu & Nelson, 1998; Richards & Allaway 1993).

5. CONCLUSIONS

It can generally be contested that there are two perspectives to explain Parasuraman`s model. Firstly, the guests directly gain loyalty from the service quality dimensions: 'tangibles, reliability, responsiveness, assurance, and empathy'. When they perceive a higher service quality, they will ultimately turn into a loyal guest of a specific service quality provided such as the F&B department, and secondly, guests may have higher satisfaction level. The application of this model by hotels can bring success for food and beverage departments to achieve high service quality which has a direct effect on satisfaction level of guests. As this paper is only conceptual, future studies may incorporate empirical data that will be collected from F&B departments in selected hotels in Jordan.

ACKNOWLEDGEMENT

This research has partially been supported by Universiti Sains Malaysia's fellowship scheme and research grant (1001/PTS/8660013).

REFERENCES

1. Al Khattab, S. S. J. (2011). Perceptions of Service Quality in Jordanian Hotels. International Journal of Business and Management, 6(7). http://dx.doi.org/10.5539/ijbm.v6n7p226

2. Al Rousan, R. M. R. (2011). Hotel service quality and customer loyalety in Jordanian hotels: A case stady of marriott hotels chain. University Saiance Malaysia Pula Pinang.

3. Antony, J. F., & Ghosh, S. (2004). Evaluating service quality in a UK hotel chain: a case study. International Journal of Contemporary Hospitality Management, 16(6), 380-384.

4. Armstrong, R. W., Mok, C., Go, F. M., & Chan, A. (1997). The importance of cross-cultural expectations in the measurement of service quality perceptions in the hotel industry. International Journal of Hospitality Management, 16(2), 181–190.

5. Babajide, O. (2011). Breakfast service quality in Restaurant Gui Hilton Helsinki Airport hotel. Bachelor's Thesis, HAAGA- HELIA University of applied sciences.

6. Carev, D. (2008). Guest satesfaction and gest loyalety study for hotel industry. College of Applied Science and Technology, Pro Quest. database.

7. Carman, J. M. (1990). Consumer perceptions of service quality: an assessment of the SERVQUAL dimensions. Journal of Retailing, 66(1), 33-55.

8. Cheung, C., & Law, R. (1998). Hospitality service quality and the role of performance appraisal. Journal of managing service quality, 8(6), 402-406.

9. Choi , T. Y., & Chu , R. K. S. (1998). Consumer perceptions of the quality of services in three hotel categories in Hong Kong. Journal of Vacation Marketing, 5(2), 176-189.

10. Christie, R. M. (2002). A Comprehensive Model of Customer Satisfaction in Hospitality and Tourism: Strategic Implications for Management. International Business &Economics Research Journal, 1(6), 8-18.

11. Crick, A. P., & Spencer, A. (2011). Hospitality quality: new directions and new challenges. International Journal of Contemporary Hospitality Management, 23(4), 463-478.

12. Cronin, J. J., & Taylor, S. A. (1992a). SERVPERF versus SURVQUAL:reconciling performance - based and perceptions - minus - expectations measurment of service quality. The Journal of Marketing, 58, 55-68.

13. Cronin, J. J., Brady, M. K., & Hult, G. T. (2000). Assessing effects of the quality, value, and customer satesfaction on consumeral behavioral intentions in service environments. Journal of Retailing, 76(2), 193-218.

14. Duncan, E., & Elliott, G. (2002). Customer service quality and financial performance among Australian retail financial institutions. Journal of Financial Services Marketing, 7(1), 25-41.

15. Getty, J. M., & Thompson, N. K. (1994). The relationship between quality, satisfaction, and recommending behaviour in lodging decisions. Journal of Hospitality and Leisure Marketing, 2, 3-22.

16. Gounaris, S. P., Stathakopoulos, V., & Athanassopoulos, A. D. (2003). Antecedents to perceived service quality: an exploratory study in the banking industry. The International Journal of Bank Marketing, 21(4/5), 168- 190.

17. Harvey, J. (1998). Service quality: a tutorial. Journal of Operations Management, 16, 583-597.

18. Hersh, A. M. (2010). Evaluate the impact of Tourism Services Quality on customer's satisfaction. Interdisciplinary journal of contemporary research in business, 2(6), 1-29.

19. Jabulani, N. (2001). An examination of customer service employees self effeicancy, job stesfaction, demographic factors, and customer perception of hotel service quality delevery in jamaica. Noava southeastrean universitey USA, ProQuest Information database.

20. Johnson, R. L., Tsirps, M., & Lancioni, R. A. (1995). Measuring service quality: A systems approach. Journal of Services Marketing, 5, 6-19.

21. Kang, G. D., & James, J. J. (2004). Service quality dimensions: an examination of Gro¨nroos's service quality model. Journal of Managing Service Quality, 14(4), 266-277.

22. Knuston, B., Stevens, P., Wullaert, C., Patton, M., & Yokoyama, F. (1991). Lodgserv: A service quality index for the lodging industry. Hospitality Research Journal, 14(3), 277-284.

23. Knutson, B., Stevens, P., Wullaert, C., Patton, M., & Yokoyama, F. (1990). LODGSERV: a service quality index for the lodging industry. Hospitality Research Journal, 14(2), 227-284.

24. Kotler, P., & Bloom, P. N. (1984). Marketing professtional services. Engleewood cliffs, NJ: Prentice - Hall. Ladhari, R. (2009). Service quality, emotional satisfaction, and behavioural intentions A study

in the hotel industry. Journal of Managing Service Quality .Emerald Group Publishing Limited, 19(3), 309-331.

25. Lam, T., & Zhang, Q. A. (1998). Study of service quality: the case of travel agents in Hong Kong. Tourism and Hotel Industry in Indo-China South East Asia. Paper presented at the Development Marketing and Sustainability, 3rd International Conference, Proceedings 4-6 June, Thailand. Lidén, S. B. (2003). Customer expectations on service guarantees. Managing Service Quality, 13(5), 338-348.

26. Mei, A. W. O., Dean, A. M., & White, C. J. (1999). Analyzing service quality in the hospitality industry. Journal of Managing Service Quality, 9(2), 136-143.

27. Mill, R. C. (2002). A Comprehensive Model Of Customer Satisfaction In Hospitality And Tourism:Strategic Implications For Management. International Business & Economics Research Journal, 1(6), 7-18.

28. Mohsin, A. (2011). Exploring Service Quality in Luxury Hotels: Case of Lahore, Pakistan. The Journal of American Academy of Business, Cambridge, 16(2), 296-303.

29. O'Neill, M. A. (2001). Measuring service quality and customer satisfaction. In J. Kandampully., C. Mok, & B. Sparks, (Eds.), Service quality management in hospitality, tourism and leisure. New York: Haworth Hospitality Press.

30. Oberoi, U., & Hales, C. (1990). Assessing the quality of the conference hotel service product:towards an empirically based model. Service Industries Journal, 10(4), 21-700.

31. Oliver. (1999). Whence Consumer Loyalty? International Journal of Marketing, 63, 33-44.

32. Parasuraman, A., Zeithaml, V. A., & Berry, L. L. (1985). A conceptual model of service quality and its implications for future research. Journal of Marketing, 49, 41-50.

33. Parasuraman., Zeithaml, V. A., & Berry, L. L. (1988a). SERVQUAL: a multiple item scale for measuring consumer perceptions of service quality. Journal of Retailing, 64(1), 12-40.

34. Peter, S. C., & Nicole, K. (2006). Swedish hotel service quality and loyality dimention. Advances in Hospitality and Leisure, 2, 123-157.

35. Pitt, L. F., & Jeantrout, B. (1994). Management of customer expectations in service firms: A study and a checklist. The Service Industries Journal, 14(2), 170-189.

36. Powapaka, S. (1996). The role of outcom quality as adeterminantof overall service quality in different categories of service industries: an empirical investigation. Journal of Services Marketing, 10(2), 5-25

37. Presbury, R., Fitzgerald, A., & Chapman, R. (2005). Impediments to improvements in service quality in luxury hotels. Managing Service Quality, 15(4), 357-373.

38. Qu, H., & Nelson, T. (1998). Service Quality Gap in China's Hotel Industry: a Study of Tourist Perceptions and Expectations. Journal of Hospitality & Tourism Research, 22(3), 252-267.

39. Rust, R. T., & Zahorik, A. J. (1993). Customer satesfaction, customer retention, and market share. Journal of Retailing, 69(2), 103-125.

40. Rust, R. T., Zahorik, A. J., & Keiningham, T. L. (1995). Return on quality (ROQ): Making service quality financally accountable. The Journal of Marketing, 59(2), 58-70.

41. Saleh, F., & Ryan, C. (1991). Analysing service quality in the hospitality industry using the SERVQUAL model. Services Industries Journal, 11(3), 324-343.

42. Soutar, G. N., Mok, C., & Sparks, B. (2001). Service quality, customer satisfaction, and value: an examination of their relation-ships in Kandampully Service quality management in hospitality, tour-ism, and leisure. New York: The Haworth Hospitality Press.

43. Su, A. Y. L. (2004). Customer satisfaction measurement practice in Taiwan hotels. Journal of Hospitality Management, 23, 397-408.

44. Sulek, J. S., & Hensley, R. L. (2010). Updating service operations. Managing Service Quality, 20(5), 475-489.

45. Tat, Y. C., & Raymound, K. S. C. (1999). Consumer perceptions of the quality of services in three hotel categories in Hong Kong. Journal of Vacation Marketing, 5(2), 176-189.

46. Upal, M. (2008). Telecommunication Service Gap: Call Center Service Quality Perception and Satisfaction. Journal of Communications of the IBIMA, 3.

47. Zeithaml, Berry, L. L., & Parasuraman, A. (1993). The Nature and Determints of customer expectations of service. Journal of the academy of Marketing service, 21(1), 1.

48. Zeithaml, Berry, L., & Parasuraman, A. (1996). The behavioural consequenses of service quality. Journal of Marketing, 60, 31-46.

49. Zeithaml., & Bitner, M. J. (2003). Services Marketing: Integrating Customer Focus Across the Firm. New York: McGraw-Hill.

6

WHEN IS A GOOD DEAL BAD FOR BUSINESS? AN EXPLORATORY STUDY OF ROOM RATES HIGHLIGHTING THE PRACTICE AND EFFECTS OF DISCOUNTING ROOM RATES

Gwen Williams

University of Nevada, Las Vegas

PART I

1. INTRODUCTION

The recession of 2008-2009, which has carried over into 2010, has generally wreaked havoc on the financial state of the hospitality industry. One example of this havoc is the fact that from the beginning to the end of 2009, hotel foreclosures in California increased by 313% and the number of hotels in California that defaulted on mortgages increased by 479% (Lansner, 2010).

Other areas in the United States such as Las Vegas, Orlando and New York City have been hit hard too. Many hotels and resorts in these popular tourist destinations as well as in other locations have adopted the practice of discounting, often deeply, their room rates for what seems like the long-term.

An example of this deep discounting is when Starwood's Resort at Singer Island in Palm Beach, Florida offered a 30% room rate discount that was marketed as their "Winter Sale" (Twelve Vacation Days Publishing, 2010). Starwood's Resort at Singer Island in Palm Beach and other hospitality industry businesses have gone to discounting in hopes of attracting more customers and thus continuing to make money, as a hospitality business must. The following decreases in ADR or RevPAR for 2009 for Las Vegas, Orlando and New York City were essentially due to hotels and resorts in these locations discounting room rates: (a) Las Vegas - 27% decrease in ADR (Lewis, 2009), (b) Orlando - 11% decrease in ADR (Orange County CVB, 2009), and (c) New York City - 26.3% decrease in RevPAR (Reed Business Information, 2010b). Is the practice of discounting working for or against hotels and resorts?

Rewind to the aftermath of September 11, 2001. The terrorist attacks on the Twin Towers in New York City paralyzed the air-traveling public and brought the hospitality/tourism industry to its knees for a relatively short time. Many hotels and resorts discounted room rates during that time too. Did discounting room rates work then? The Center for Hospitality Research at Cornell University published a very interesting and telling study done on the practice of discounting room rates revolving around the effects September 11, 2001 had on the hospitality/tourism industry (Enz, Camina and Lomanno, 2004).

One method for setting room rates sees them set based on a variety of factors including, but not limited to: (a) competitor's room rates, (b) cost of running business and (c) customer Comfort Level (Waller, 1997). By discounting the room rates that have been set based on the aforementioned factors in addition to any other factors that come into play, can the cost of doing business still be covered? Are the discounted room rates below what the competitors are getting for their rooms? Is the discounted room rate lower than what customers are willing to pay for a room? Those are just a few questions a savvy hotelier could ask about discounting.

Other questions that might arise out of the practice of discounting room rates include: (a) What about rate integrity?, (b) What are the financial effects of discounting?, (c) What are the effects, aside from financial ones, of discounting?, (d) What about brand integrity?, (e) Are there some hotels and resorts that should not give room rate discounts?, and (f) If hotels and resorts truly feel the need to discount - What would be a good formula that could be used to make sure too deep of discounts are not given?

Through a literature review on this paper's topics, the reader will discover the answers to the above listed questions. The reader will also gain insight into some methods for setting room rates as well as get a more in-depth look at the practice of discounting room rates for the long-term and the consequences that can occur because of it.

1.1 Purpose

The purpose of this paper is two-fold: To suggest who should and who should not discount room rates and to convey methods that can be used for discounting room rates without loss of a target market and/or without financial loss. To arrive at this paper's purpose, this paper will explore and analyze: (a) methods used for setting room rates, (b) setting room rates for profit maximization, (c) rate and brand integrity, (d) methods used for discounting room rates, and (e) the positive and negative effects of discounting room rates.

1.2 Problem Statement

Many hotels and resorts are discounting their room rates in an attempt to cope with the current financial crisis and economic recession. However, the practice of discounting room rates can easily lead to the loss of a target market and/or to financial loss in both the short-term and the long-term. Whether or not hotels and resorts should discount their room rates and if they discount how they should discount their room rates are issues that urgently need investigation.

1.3 Justification

The economy's downward spiral that started at the end of 2008 has stuck around and has caused: (a) high unemployment rates (U.S. Bureau of Labor Statistics, n.d.); (b) lower wages (Shedlock, 2009), (c) record hotel foreclosures (Lansner, 2010), (d) bank collapse of Washington Mutual – one of the largest banks in the U.S. (Russell, 2009); and (e) a sizeable drop in business and leisure travel (Office of Travel and Tourism Industries, n.d.). In the wake of this economic turmoil, owners of hotels and resorts are trying to maintain or even to increase their occupancy rates. To do so, many began to offer room rate discounts – some as much as 27% lower than their ADR (Lewis, 2009).

According to Smith Travel Research, the practice of discounting decreased the U.S. hotel industry's RevPAR by 16.7% for the year 2009 (Reed Business Information, 2010b). President of Smith Travel Research, Mark Lomanno, commented that 2009 would be remembered as the worst year in modern hotel industry (Reed Business Information, 2010b).

There are many factors that should be taken into consideration when setting, and especially when discounting, room rates. Discounting room rates is usually done when a hotel or resort feels the need to realize a higher occupancy rate. The discounts that are being offered may indeed not be in the best interest of the hotel or resort that is giving the discount (Feiertag, 1993).

It is important for the owner and for the managers of a hotel or resort to understand the ramifications of discounting room rates (Feiertag, 1993). In turn, it would be helpful for the owner and the managers to have a formula to go to when calculating room rate discounts if they really feel the need to use the practice of discounting (Smith, 2009).

1.4 Constraints

One of the limitations of this paper is the fact that it does not speak specifically to a certain hotel or resort but rather generally to all. This is because hotels and resorts are unwilling to publish or to share their discounting data as it is regarded as confidential corporate information.

A second limitation of this paper is that its presentation of methods used for setting and/or discounting room rates is not exhaustive. The methods presented are what were available through research of this paper's topics.

A third limitation of this paper is that some of the methods presented for setting room rates and for discounting room rates, though they represent what has been used or what is being used in the industry, may not work for every hotel or resort. In the end, it is up to the individual property's ownership and management to decide which method will work best for their property.

1.5 Glossary

Following are definitions for some of the core terms and concepts used throughout this paper. These core terms and concepts are explained in more detail in the body of the paper.

ADR – Average Daily Rate

Hotel – Provides lodging with limited amenities and limited or no personal services.

Resort - Provides lodging, a wide array of amenities and a wide array of personal services.

Revenue Management – Method used to set room rates that maximizes revenue and manages demand by using pricing and duration controls. The "new" Yield Management.

RevPAR – Revenue Per Available Room

Yield Management – Method used for setting room rates to maximize room revenue and profits.

PART II

2. LITERATURE REVIEW

2.1 Introduction

The current economic recession that began in 2008 has, once again, brought about the practice of discounting room rates – often deeply (Lewis, 2009). Historically, another time of economic slow down for the hospitality industry also brought about the same practice. This time began with the recession of 2001 that was magnified after September 11, 2001 (Lewis, 2009).

Enz et al. (2004) released a report that they wrote based on a study performed about the years 2001-2003. The study revolved around the practice of room rate discounting that occurred during that time period's economic downturn for the hospitality industry. The study was based on hotels and resorts in a competitive market. To reaffirm the study's findings, Canina and Enz (2006) released a follow up report based on a study done in 2004 when the economy stabilized. This study reported the same findings as the 2001-2003 study did.

With the assistance of the aforementioned literature as well as other topical literature, the Literature Review section of this paper will explore the nuances of setting room rates, address rate as well as brand integrity and end with a discussion on the practice of discounting room rates including the positive and negative effects that come from this practice.

2.2 Methods used for setting room rates

The methods for setting room rates that will be presented in this paper are broken down into the three following categories: (a) based on cost, (b) based on market, and (c) based on combination of cost and market.

2.2.1 Methods Based on Cost

Per Schmidgall (2006), the two most common methods for setting room rates that are based on cost are the $1 for every $1,000 method and the Hubbart Formula. The $1 for every $1,000 method makes each room cost $1 for every $1,000 of construction cost. For example, a 300-room hotel is built for a price of $40 million or a per room cost of $133,333.33 per room. Therefore, the room rate for a room would be set at $133.33 per night if using the $1 for every $1,000 method to set room rates. As Schmidgall (2006) noted, this method does not take the current value of the building into consideration. This method also does not take the other services the hotel may offer such as a restaurant, spa and bar into consideration. In addition, it ignores current operating costs.

The Hubbart Formula is a procedure that sets the average room rate through the use of eight steps (Steed and Gu, 2005). Schmidgall (2006) commented that the Hubbart Formula starts with the property's preferred profit. The eight steps to the Hubbart Formula as laid out by Schmidgall (2006) are:

1. Step One – Compute the property's preferred profit by multiplying the wanted return on investment by the investment of the owner.

2. Step Two – Compute before tax profits by dividing the number arrived at in

Step One (property's preferred profit) by the tax rate subtracted from 1.

3. Step Three – Compute fixed costs and any management fees.

4. Step Four – Compute undistributed operating costs.

5. Step Five – Compute departmental operating costs leaving out the Rooms department.

6. Step Six – Compute Rooms department income by adding Steps 2,3,4 and 5 together.

7. Step Seven – Compute the revenue for the Rooms department by adding all Rooms department costs to Step 6.

8. Step Eight – Compute the average room rate divided by the number of rooms expected to be sold.

As noted by Steed and Gu (2005), the $1 for every $1000 method and the Hubbart Formula have been discarded by today's hotels and resorts because these methods for setting room rates that are based on cost do not take other aspects that should affect a property's ADR into consideration. Waller (1997) suggested that in addition to operating costs, a property's competitor's room prices as well as how much a property's customers are willing to pay for a property's rooms should be taken into consideration when setting the ADR.

2.2.2 Methods Based on Market

The methods based on market for determining room rates that will be discussed here are: (a) competitor-based pricing, (b) behavioral-based pricing, and (c) yield management.

Competitor-based pricing. One method for setting room rates based on market can be called competitor-based pricing - it is also known as "going-rate pricing" (Kotler et al., 2010). Going-rate pricing is when pricing is set according to a property's competitor's prices. In a market whose elasticity is hard to determine, going-rate pricing is a good method to use for setting room rates (Kotler et al., 2010). It is a good method because it feeds off of the group intellect of a property's competitors and may, also, avoid "price wars" (Kotler et al., p. 306, 2010).

Before pricing off competitors, a property must first research who its competitors are by performing a competitive analysis (Kotler et al, 2010). As recorded by Kotler et al. (2010), a successful competitive analysis will take more factors than just the physical ones into consideration. These other factors include service standards and general property & room cleanliness.

When a property has found its true competitors, that property's room rates can be set off of the room rates of those competitors. Enz et al. (2004) reported that to attain and/or to maintain a respectable RevPAR against a competitor's a property should set their room rates using + or - 2% as a guide. In other words, a property that is setting their room rates off of their competitor's room rates should not go lower or higher than 2% from their competitor's room rates.

In the study that was performed by Enz et al. (2004), properties that set their room rates greater than 2% below that of their competitors may have realized more occupancy than their competitors however their overall revenue was lower in comparison. On the other end of the scale, a property that set their room rates 2% or more higher than that of their competitors realized higher revenue in comparison (Enz et al., 2004) even when their occupancy rate was not as high.

Behavioral-based pricing. Another method for setting room rates that is based on market can be called behavioral pricing - it is also known as "value-based pricing" (Kotler et al., p. 305, 2010). This method takes Waller's (1997) suggestion of a property knowing its "Customer Comfort Zone" to heart. In order to use this method, a hotel or resort will need to perform some market research. This research can be done in a variety of ways however needs to ensure that all market segments are surveyed properly (Kotler et al., 2010). A simple strategy for polling market segments' value perceptions would be to attach the survey to a property's guest satisfaction survey with a question similar to "At what price would this room be too expensive?" (Waller, 1997).

Once a hotel or resort has gathered and assessed all of the value perceptions from its market segments, the property can set room rates with this information in mind. The business segment may have higher room rates than the leisure segment or vice versa (Kotler et al., 2010). Steed et al. (2005) commented that the value-based pricing method works very well with the group market segment. Economy hotels are another good outlet for the use of this method of setting room rates as economy hotel guests are quite price sensitive (Enz, Canina and Lomanno, 2009).

Yield management. To set room rates based on a property's market, Waller (1997) advised to track market demand of a property as well as market demand of its competitors while anticipating any supply changes. Waller (1997), also, commented that a property should look at their room pricing strategy as it relates to their market share instead of how it relates to their occupancy percentages. In other words, if a property enjoys a year-round occupancy percentage that hovers between 85% and 95% that property should look more to the seasonality of the market as opposed to their occupancy percentages to set room rates more accurately. Yield management takes market trends including seasonality into account (Relihan, 1989).

In essence, yield management endeavors to set room rates in line with market trends (Relihan, 1989). Therefore, yield management is another method that is used to set room rates that is based on market. To use yield management for setting room rates, a hotel or resort must comprehend their guests' buying behaviors and draw comparisons between current and future room demands (Relihan, 1989). Using yield management to set room rates can help a hotel or resort realize equilibrium for supply and demand. This is because it will allow a hotel or resort to make small room rate adjustments in anticipation of seasonality and other market trends. According to Withiam (p. 4, 2001), "...yield management uses the basic strategy of providing the right service at the right time to the right customer at the right price."

The room rate fluctuations that can be seen from a hotel or resort practicing yield management should not be confused with room rate fluctuations that can be seen from a hotel or resort that is simply discounting their room rates. In its function as a method for setting room rates, yield management sets rate fences for discounted prices (Withiam, 2001). Rate fences are limits on discounts that can be given on room rates.

To put yield management into practice for setting room rates, Withiam (p.4, 2001) suggested remembering the "4-c's: calendar, clock, capacity and cost". Calendar refers to managing the date of the guests' reservation. Clock refers to the timing of the guests' stay. Capacity refers to finding equilibrium between supply and demand. Cost refers to the room rate that is set in relation to calendar, clock and capacity.

The basis for a good hotel or resort yield management program is the ability to project room demand (Withiam, 2001). To project demand accurately, the demand for each market segment or rate class must be known. The seasonality (timing of demand) must also be known. To further project and manage demand a hotel or resort must track the length of guest stays. This will assist the hotel or resort in controlling capacity. Once demand, inclusive of seasonality, is accurately projected and capacity is controlled, a hotel or resort can go about setting room rates. Setting room rates is the most difficult task of practicing yield management for a hotel or resort (Withiam, 2001).

When setting the different price categories, a hotel or resort should utilize strong rate fences. Strong rate fences will keep guests from getting too large of a discount (Withiam, 2001). Rate fences can be physical like the location of a room, the size of a room or the in-room amenities. Rate fences can also be related to guests' group memberships such as AAA or AARP. Other rate fences include days of the week.

The largest possible downfall of yield management happens when a guest is upset by feeling that a hotel's or a resort's pricing practices are unfair because that guest did not qualify for a lower room rate due to the rate fences a hotel or resort has set (Withiam, 2001). This downfall can be circumvented by a hotel or resort by ensuring that guests have knowledge of that hotel's or of that resort's pricing practices (Taylor and Kimes, 2010). As laid out by Taylor et al. (2010), to ensure that guests have knowledge of a hotel or resort's pricing practices that hotel or resort can: (a) make information available on the property's website, third-party websites and through the reservations department regarding the different room rates and the reasons for the differences; (b) advertise any restrictions along with special room rates or packages; and (c) make sure reservation and front desk employees have knowledge of the different rates and the rate fences that go along with each rate so they can explain them accurately to guests. When guests feel that they are being treated fairly, the likelihood of a return stay rises exponentially (Taylor et al., 2010). Using yield management for profit maximization will be covered in more detail later in this paper.

2.2.3 Methods based on a combination of cost and market

Kotler, Bowen and Makens (2010) conceded that costs set the lower price limits for setting room rates but that it is the market and demand that set the upper price limits for setting room rates. Per Pan (p. 750, 2007), seasonality definitely affects room rates in both "low and high" seasons. "Low" season occurs when a hospitality property is experiencing a low rate of visitation. "High" season occurs when a hospitality property is experiencing a high rate of visitation.

Daily variable costs per room sold and room capacity also have effects on room rates. However, as recorded by Pan (2007), the daily variable costs per room sold only affect room rates set during the low season. Because room capacity has the greatest impact on room rates set during the high season, it is a negative impact.

To account for the daily variable costs, Pan (p. 750, 2007) proposed the following two models for setting room rates, which are based on a combination of cost and market and take into consideration not only the daily variable costs but also seasonality (market demand) and room capacity:

Low season average room rate (P^{L*}) = $(A^L + r)^{/2}$

High season average room rate (P^{H*}) = $A^H - Q$

Where P^{L*} is the average room rate set for low season, A^L is the room demand in low season when rooms are free, r is the variable cost per room sold, P^{H*} is the average room rate set for high season, A^H is the room demand in high season when rooms are free, and Q is the total number of rooms.

Pan (2007) based his models on the hypotheses that (a) higher high season demand will increase the high season room rate without affecting the low season room rate; (b) higher low season demand will increase the low season room rate without affecting the high season room rate; (c) if the highest demand in the high season, the highest demand in the low season and the daily variable costs per each room sold remain the same then an addition to a property's room capacity will cause a decrease in the equation $P^{H*} - P^{L*}$; and (d) if the daily variable costs per room sold go up when the greatest demand in the high season, the greatest demand in the low season, and the room capacity stay the same then there will be a decrease in the equation $P^{H*} - P^{L*}$. When Pan (2007) developed a regression model to test these hypotheses using data from hotels and resorts in Taipei, Taiwan, he found that his equations provided the best means of setting room rates for the high and low seasons.

The basis for Pan's (2007) model was Gu's (1997) model for setting room rates based on a combination of cost and market which is expressed in the following formula:

$$r^* = (a + ßv)/2ß$$

where r* is the optimal room rate, a is the possible room demand when rooms are free, ß is the measure for demand that is price sensitive, and v is the variable costs per rented room.

Because this model accounts for the demand for rooms, as well as the variable costs per rented room, it is another method for setting room rates that is based on the combination of cost and market factors. This model requires that demand characteristics and variable costs per rented room need are recalculated often to ensure that the model is continuously adjusted to the factors that influence it.

In the multi-stage synthetic hotel room pricing model, a third method that can be used of setting room rates based on cost and market factors, Kim, Han and Hyun (2004) combined the methods for setting room rates based on cost with the methods for setting room rates based on market according to the following seven steps:

1) Identify target markets, market position and competitors for the same markets. Identify competitor's average price (room rate).

2) Calculate "Cost-Based Price (CBP): CBP = (VC+FC+PG)" (Kim et al., p. 172, 2004).

3) Determine the property's "Differentiation Premium (DP)" (Kim et al., p. 172, 2004). Compute "Market Premium Price (MPP): MPP= (1+DP) x Average competitor's price" (Kim et al., p. 172, 2004).

4) Contrast MPP with CBP. If "MPP>CBP" (Kim et al., p.172, 2004) then use MPP as price and go to next step, if MPP

5) Ensure price set (MPP) is within limits between the pricing ceiling and pricing floor.

6) Apply rate fences to price.

7) Apply seasonality to price.

2.2.4 Setting room rates for Profit Maximization

According to Butscher,Vidal and Dimier (2009), the ability to set room rates is the best tool that hotels and resorts have to optimize their revenues, even during economic downturns. As a goal of a good yield management program is to maximize profits (Relihan, 1989), this section discusses the use of yield management as a way for hotels and resorts to maximize profits through setting the appropriate room rates.

2.2.5 Yield Management

When practicing yield management for profit maximization, hotels and resorts must control room rates to avoid large differences among different rates that could upset guests (Relihan, 1989). Gu and Caneen (1998) presented the following two quadratic models to help maximize profits through setting the appropriate room rates. These models do not help hotels and resorts determine room rates but rather calculate how to optimize revenue from setting the appropriate room rates.

Ym quadratic formula I. According to Gu and Caneen's (1998) first quadratic formula for maximizing revenue from room rates,

$$r^* = a/2ß$$

where r* is the optimal room rate, a is the room demand when the ADR is zero and ß measures price sensitivity of demand

Gu and Caneen (1998) explained that, "If the actual ADR is higher than the model-calculated r*, reducing ADR may increase room revenue. If the actual ADR is lower than r*, raising room rates could lead to higher room revenue" (Gu et al., p. 249, 1998). Thus, hotels and resorts can use this equation as a means of determining when to make an adjustment to room rates.

Ym quadratic formula II. According to Gu and Caneen's (1998) second quadratic formula, which incorporates cost considerations,

$$r^* = (a + ßv)/2ß$$

where r* is the optimal room rate, a is the room demand when the ADR is zero, ß measures price sensitivity of demand and v is the variable cost per rented room.

2.2.6 Revenue Management

Revenue management is the hospitality industry's new yield management. Although the terms are often used interchangeably, there has been a definite transition from yield management to revenue management in the hospitality industry (Cross, Higbie and Cross, 2009). While yield management was more concentrated on inventory, revenue management has become more concentrated on the guests. Instead of just managing demand, the modern practice of revenue management has an important role in the creation of demand, making it imperative for a revenue management department to work closely with its sales and marketing departments. When the departments all work together on pricing, revenue management can truly have a positive effect on all aspects of a hotel's or resort's business.

Cross et al. (2009) contended that there are four main areas of concentration that have emerged from the transition from a focus on inventory to a focus on guests: (a) to maximize price, not just inventory; (b) to find solid pricing strategies that will enable multiple rates each day of the year; (c) to have a comprehension of guests' price comfort level; and (d) to control positioning in the marketplace. The process of setting room rates using revenue management, which is almost identical to setting room rates using yield management, includes: (a) segmenting customers by market, (b) ensuring rates for each market segment do not contradict one another, (c) projecting seasonality, and (d) limiting practice of overbooking (Vinod, 2004). To truly maximize profits from room rates, today's revenue management can take advantage of sophisticated software (Cross et al., 2009). The appropriate information is input into the program which then calculates room rates that will optimize revenue.

Although RevPAR is a strong indicator of the effect of revenue management on the entire business of a hotel or resort, an even better indicator is the Revenue Per Available Customer (Cross et al., 2009). By consistently measuring the Revenue Per Available Customer, a hotel or resort can better track spending by market segment, allowing it to maximize pricing even more throughout its property.

Revenueyourhotel.com and the HSMAI Revenue Management Advisory Board (n.d.), laid out the following action items for all revenue managers for 2010:

1) Ensure everyone involved in owning and operating the hotel or resort has knowledge of the goals of revenue management.

2) Recover rates as quickly as possible.

3) Gain a better comprehension of market segments so solid forecasts can be made.

4) Collect the necessary data needed for strong pricing decisions.

5) Make revenue management a part of managing the guest relationship with the hotel or resort.

6) Become more strategy oriented instead of just being analytical.

7) Gain more insight into marketing on the internet.

8) Utilize software and other available revenue management tools to enable better forecasting.

9) Concentrate on increasing Revenue Per Available Customer.

10) Communicate focus on customer value instead of on price.

2.2.7 Rate and Brand Integrity

Rate Integrity in the hospitality industry can be defined as maintaining "...general prices based on the value of the product..." (Martinez, p. 19, 2009). Rate Integrity is gauged by comparing a property's different room rates (Wolfe and Boger, 2001). Low rate integrity occurs when there are large differences among rates. Rate Integrity is important for hotels and resorts because it keeps prices consistent with the guests' perceived value of the property, thereby decreasing their likelihood of being sued over pricing practices considered unfair.

Brand Integrity in the hospitality industry can be defined as "...a desirable outcome for companies seeking to maintain a consistent, unambiguous position in the mind of their audience" (Wikipedia, 2010), with the audience in the case of hotels and resorts being the guests. Much of a hotel or resort's brand integrity rests not only on physical accommodations but also on the number and quality of services provided by its employees (Mangan and Collins, 2002). As rate integrity and brand integrity together determine guests' loyalty to a hotel or resort, any fluctuation in either factor can lead guests to question their loyalty (Mangan et al., 2002).

2.2.8 Methods used for discounting room rates

An effect of the current recession is that many people have become more pricesensitive (Duff, 2009). By offering room rate discounts, hotels and resorts in the hospitality industry are hoping to appeal to people's price-sensitivity.

2.2.9 Housekeeping discounts

One method for discounting room rates used by Starwood Hotels & Resorts Worldwide, Inc. is to extend hotel credit or points in its guest loyalty program to guests who choose not to have housekeeping services for their rooms (Nassauer, 2010). For each night that a guest does not require housekeeping, Starwood gives a $5 hotel credit or 500 points towards the guest loyalty program. Other hotels that use this method for discounting include Marmara Manhattan, which gives guests a $20 discount for every night they do not require housekeeping (Nassauer, 2010), and Marriott's Grand Chateau in Las Vegas (M. Colvin, personal communication, February 12, 2010). This method of discounting rooms has not cost for the hotel – in fact, it saves the hotel the costs associated with cleaning a room - and gives the guest the option of taking the discount by not requiring housekeeping services on a daily basis (Nassauer, 2010).

2.2.10 Competitor-Based discounts

Another method used for discounting room rates in a hotel is similar to the competitor-based method for setting room rates. If a hotel's competitors are giving room rate discounts, a hotel may feel pressured to do the same by either matching their competitor's discounts or by giving an even deeper discount (Enz et al., 2004).

2.2.11 Discount packages

A third method resorts use for discounting room rates is a method referred to as packaging or bundling (Kotler et al., 2010). One example of this is Chula Vista Resort's All-Inclusive package which includes the room itself, all food and drinks, and unlimited waterpark access for $199 for two nights, not including weekends (other restrictions apply; Chula Vista Resort, 2010). The room rate outside of the inclusive package was $99 per night.

2.2.12 Discount based on historical data

A fourth method used for discounting room rates is to determine the discount based on historical data (R. Roger, personal communication, March 2, 2010). If a certain room rate filled rooms in the past, it is used to fill rooms again, even if doing so requires a 50% discount off the current ADR.

2.2.13 Discount based on promotions

One more method often used for discounting is to run promotions. One popular promotion is offering a free night after a stay of a certain number of nights. Examples of this type of discounting have been used at Starwood's Resort at Singer Island in Palm Beach, FL (Twelve Vacation Days Publishing, 2009), the Grand Hotel Marriott Resort, Golf Club & Spa in Point Clear, AL (Marriott International, Inc., 2010) and the Four Seasons Hotel in Las Vegas (Four Seasons Hotel, 2010).

2.3 Positive Effects of Discounting Room Rates

2.3.1 Higher occupancy levels

Higher occupancy levels are a possible positive effect of discounting room rates. According to the study performed by Enz et al. (2004), hotels and resorts in a competitive market did indeed experience slightly higher occupancy levels when they discounted their room rates.

2.3.2 Gaining market share

Gaining market share has been another possible positive effect of discounting room rates. Hotels and resorts in a competitive market may be able to grab some market share from a competitor by discounting their room rates before their competitors match the discount (Lomanno, 2008).

2.4 Negative Effects of Discounting Room Rates

2.4.1 Decreased Rate Integrity

According to research done by Kimes (2009), it is difficult for revenue managers facing an economic downturn to maintain rate integrity because it is difficult to convince their owners to hold rates without discounting to protect rate integrity. Lomanno (2008) found that by offering discounts, hotels and resorts are conditioning their guests to expect the discounted prices even after the recession lifts, requiring years to pass before owners can return rates back to their original pre-discount period levels. For example, six years passed after September 11, 2001 before hotels and resorts in New York City could return rates to their pre-9/11 prices (Lomanno, 2008).

2.4.2 Decreased Brand Integrity

Kimes (2009) also found that revenue managers are concerned that discounting can damage brand integrity. In many cases, the larger the discount given, the greater the potential for damage was to a property's brand integrity. Lomanno (2008) supported revenue managers' concern by contending that by discounting room rates, hotels and resorts teach their guests that a different value has been set for that property.

2.4.3 Price Wars

Kimes (2009) reported that surveyed revenue managers were also concerned about the price wars that discounting has produced. In the face of price wars, revenue managers felt great pressure to discount rates in order to maintain and compete for market share. Unfortunately, these price wars were over who could charge the lowest price for a room.

2.4.4 Lower RevPAR

In a competitive market, hotels and resorts whose rates were discounted more than 2% below that of their competitors had a lower RevPAR than their competitors (Canina and Enz, 2006). Canina et al. (2006) found a 4.44% lower RevPAR as compared to competitors when rates were 12% to 15% lower than competitor's rates and a 2.51% lower RevPAR as compared to competitors when rates were 6% to 8% lower than competitor's rates.

2.4.5 Decreased service quality

Revenue for many hotels and resorts has dropped due, in part, to the discounts many hotels and resorts have been giving (Reed Business Information, 2010a). To make up for the lost revenue, many hotels and resorts have laid off employees and, having fewer employees to service their guests, have experienced a decrease in service quality that, as reported by Reed Business Information (2010a), has not gone unnoticed by guests.

2.4.6 Loss of target market

Discounted room rates tend to attract a market segment that is different from a hotel's or resort's typical segment. For Chula Vista Resort in Wisconsin Dells, Wisconsin, this market is more price-sensitive, more spendthrift and less respectful of physical property than its usual target market, which has led to an increase in guest complaints and property damage (P. Finnegan, personal communication, February 20, 2010).

3. CONCLUSION

As discussed in the literature review, many practices used for setting room rates and optimizing profit from those room rates are widely available to hotels and resorts. However, once hotels and resorts start discounting room rates, all practices used for setting room rates and for maximizing revenue from those room rates are set aside.

The advantages of discounting room rates are limited to achieving slightly higher occupancy levels and temporarily gaining market share. The disadvantages to discounting room rates are many – erosion of rate and brand integrity, creation of price wars, decreasing a property's RevPAR, decreasing a property's service quality and losing an original target market.

Whether or not hotels and resorts should discount their room rates and if they discount how they should discount their room rates are issues that need answers. The answers to these issues will help mitigate the negative effects of discounting.

PART III

4. INTRODUCTION

As previously discussed, the practice of discounting room rates has more possible negative effects than positive effects on the properties offering such discounts. The answer to the question of whether or not hotels and resorts should discount their room rates is in this section of this paper. In addition, the answer to the question of if hotels and resorts are going to offer room rate discounts – how they should discount their room rates is presented in this section of this paper.

This section goes on to remind the reader of the limitations of this paper. Additionally, this section covers financial and management implications and recommendations regarding discounting room rates.

5. ANALYSIS AND RESULTS

5.1 Does discounting room rates work?

Any hotelier may be interested to know that discounting room rates is not as beneficial to a hotel or resort as it would seem. As was found by the Enz et al. (2004) study and the Canina et al. (2006) study, discounting does not work in favor of the hotel or resort that is giving room rate discounts. This was found to be especially true in a competitive market due to the availability of making comparisons. As mentioned earlier, though discounting may benefit a hotel or resort with a slightly higher occupancy level and a temporary gain in market share when compared to a non-discounting competitor it does not increase overall revenue. Discounting also erodes brand and rate integrity. The creation of price wars, a decrease in service quality and loss of an original target market are additional products of discounting room rates. Overall, discounting room rates does not always work in favor of the property doing the discounting as one would think it would.

5.2 Who should discount room rates?

Higley (2003) commented that as long as discounting is done as a part of a successful revenue management program it is an acceptable practice for that property. As previously discussed, revenue management takes many different factors into consideration when setting room rates including but not limited to customer segmentation, competitor's prices and seasonality. In agreement with Higley (2003), if a property is practicing a successful revenue management program, discounting room rates may not be harmful, as a successful revenue management program would use strong rate fences to ensure that room rates would not be too deeply discounted against the ADR or against the room rates of competitors.

An example of a successful revenue management program can be found in the Hilton Hotel Corporation. In preparation for becoming Directors of Revenue Management, Hilton sends its revenue managers to the Hilton Revenue Management University. The two-week long training program includes classes in forecasting, assessing competition, pricing strategies and other processes for making decisions (McCann, 2000). The success of Hilton's revenue management program became clear after Hilton's purchase of Doubletree Hotels. Hilton's revenue management program enabled RevPAR at Doubletree Hotels to grow from 94.1% to 99% (Center for Effective Performance, 2004).

5.3 Who should not discount room rates?

In contrast, hotels and resorts that do not have successful revenue management programs should not discount their room rates. This is because that while revenue management is very enabling of revenue maximization during good economic times, the practices that make up revenue management are even more important in a sluggish economy (Loyd-Jones, 2009). In a sluggish economy, each guest and each dollar count more than they do in a prosperous economy. A property that does not have a successful revenue management program in place exposes itself to more negative effects from an economic downturn than a property that is practicing a successful revenue management program does (Lloyd-Jones, 2009). Examples of unsuccessful revenue management programs can be found among the many hotels that foreclosed in California in 2009 (Lansner, 2010).

As was reported by the Enz et al. (2004) and Canina et al. (2006), hotels and resorts that are part of a competitive market should not discount room rates. In a competitive market, hotels and resorts that price room rates below those of their competitors realize a lower RevPAR than that of their competitors.

Hotels and resorts that are respected brands in their marketplace and seek longterm increases in room rates should also think twice before discounting room rates. As previously discussed, discounting room rates teaches guests a different value for the property, conditioning them to expect lower rates even after a recession lifts (Lomanno, 2008). For example, a guest who becomes accustomed to paying $99 for a room will balk at paying $129 for that same room under similar circumstances at a later date.

If, despite the disadvantages of doing so, a hotel or resort decides to discount room rates, it should follow the recommendations presented in the following section to continue to operate without a loss while offering the discount.

5.4 Calculating room rate discounts that do not incur losses

The following methods may be used to discount room rates without incurring loss:

1. Method 1: Calculate the breakeven point for each room type by dividing fixed costs by selling price minus variable costs (Kotler et al., 2010). The breakeven point for each room type should be the price floor. Therefore, discounts that bring room rates lower than the price floor or room rates that are below the breakeven point for each room type should not be offered.

2. Method 2: First, categorize the room types of competitors that are similar to the room types of the hotel or resort, then calculate the average room rates for market competitors per the room type

categories. Next, multiply each average room rate by at least 2%, then add this figure to the average room rates per room category. By charging at least 2% more than competitors, the RevPAR of a hotel or resort should remain respectable for that market.

3. Method 3: Use both Method 1 and Method 2.

4. Method 4: Do not discount room rates and use value-added items to add more value for guests (Nevistas and Raff, 2009). An example of a value-added item is the inclusion of a hotel or resort gift card as part of the room rate.

5. Method 5: Offer a housekeeping discount program whereby a hotel or resort extends hotel credits or points toward its guest loyalty program for guests who choose not to have housekeeping service for their room. This method for discounting rooms has not costs – it actually saves the hotel or resort the costs associated with cleaning a room - and gives the guest the option of taking the discount by not requiring housekeeping services on a daily basis (Nassauer, 2010).

5.5 Paper Limitations

The limitations of this paper are:

1. Because hotels and resorts are unwilling to publish or to share data regarding their discounting practices, this paper could only discuss the practices in general, and was unable to provide information regarding specific practices.

2. The methods presented by this paper for setting and discounting room rates are not inclusive of every method that may exist. The methods included are what were available through the research of literature on the topics of setting and discounting room rates.

3. The methods presented by this paper for setting and discounting room rates are not inclusive of every method that may exist. The methods discussed are those presented in the literature pertaining to the topics of setting and discounting room rates at the time of the research.

5.6 Financial and Managerial Implications and Recommendations

Surviving a global economic recession is not an easy task especially in an industry that relies upon leisure and business travelers. Vacations and travel for business are often the first items to get cut short or to get cut completely from personal and business budgets.

Although discounting room rates may maintain occupancy rates during an economic recession. Though discounting room rates may keep occupancy up, it does not necessarily maintain revenue. Table 1* shown below exemplifies how much more occupancy is needed to maintain the revenue level of pre-discounted pricing. For example, if present occupancy is 76% and a 5% discount in room rates is given – the property will need to achieve an occupancy of 81.4% to realize the same revenue as it did prior to giving a 5% discount in room rates.

Hotels and resorts exist for people to remain comfortable away from home, whether they are away due to leisure or business travel. During these trying economic times, it is imperative for hotels and resorts to maintain their value to guests. One way to maintain value is to maintain a level of service quality that is commensurate with the property. Another way is to keep room rates consistent with small yearly upward increments (Lomanno, 2008). Too much of a room rate discount can devalue a property from a guests' perspective.

Studies have shown that one financial implication of discounting room rates is a lower RevPAR than a competitors' RevPAR when room rates are 2% or more below that of market competitors (Canina et al., 2004). The erosion of rate integrity is another financial implication as are price wars.

Table 1. Increase in Occupancy Required to Compensate for Discount

	The price of rate-cutting				
	Reduction in present rate				
	5%	10%	15%	20%	25%
Present Occupancy	**Occupancy required to make up for reduction**				
76%	81.4%	87.7%	95.0%	103.60%	114.0%
74	79.3	85.4	92.5	100.9	111.0
72	77.1	83.1	90.0	98.2	108.0
70	75.0	80.8	87.5	95.5	105.0
68	72.9	78.5	85.0	92.7	102.0
66	70.7	76.2	82.5	90.0	99.0
64	68.6	73.8	80.0	87.3	96.0
62	66.4	71.5	77.5	84.5	93.0
60	64.3	69.2	75.0	81.8	90.0
58	62.1	66.9	72.5	79.1	87.0
56	60.0	64.6	70.0	76.4	84.0
54	57.9	62.3	67.5	73.6	81.0
52	55.7	60.0	65.0	70.9	78.0
50	53.6	57.7	62.5	68.2	75.0

(Based on cost of operating additional occupied rooms equal to 25% of present rate.)
* Source: The Horwath Accountant 47, no.7 (1967): 8.

Regarding the managerial implications of discounting, management must remain aware of all the possible financial impacts that discounting room rates can have on a property. In addition, management needs to remain cognitive of guests' perceptions of service quality especially during times of discounting. Management must also consider that a loss of an original target market can be a very real effect of discounting room rates. In other words, prior to instituting the practice of discounting room rates management must equally consider all effects of discounting room rates.

Based on the research presented in the literature review, this paper recommends that hotels and resorts only discount room rates if they have a successful revenue management program in place. The practice of discounting room rates simply has too many possible disadvantages to make it a viable solution in a sluggish economy. Instead f simply discounting room rates, alternatives, such as value-added items, can be found that will create demand and help guests place value on a property.

5.7 CONCLUSION

This paper explored the methods that hotels and resorts can use for setting and discounting room rates while maintaining rate integrity and brand integrity. Based on this exploration, it discussed the possible positive and negative effects of discounting room rates and provided suggestions regarding which methods to use if a property decides to practice discounting. Finally, this paper recommended against the practice of discounting room rates unless a hotel or resort already has a successful revenue management program in place.

Further research is needed to discover how the hospitality industry can recover from the practice of discounting room rates more quickly than history dictates. In support, Smith (2009) recommended that an industry-wide analysis of pricing be performed at the end of the current recession. This would provide the hospitality industry a more comprehensive view of pricing during an economic downturn, allowing it to better prepare for the next economic downturn.

The current recession is beginning to lift as reported by STR, who, estimates 2010 rates dropping only 3.4% (Smith, 2009). This drop will not be as significant as the 9.7% drop in rates during 2009. This paper's advice to hoteliers who did discount rates without a successful revenue management program in place is to work on quickly recovering rates back to those rates charged at the beginning of 2008.

REFERENCES

1. Butscher, S.A., Vidal, D. & Dimier, C. (2009). Managing hotels in the downturn: smart revenue growth through pricing optimization. Journal of Revenue and Pricing Management, 8(5), 405- 409.

2. Canina, L. & Enz, C. (2006). Discounting still doesn't work: a hotel pricing update. CHR Reports, 6 (2), 4-20.

3. Center for Effective Performance (2004), CEP increases market share and revenue at hilton hotels. CEP Performance Edge. Retrieved on April 4, 2010

4. Chula Vista Resort (2010). Current Specials. Retrieved March 1, 2010, from http://www.chulavistaresort.com/specials-packages/current-specials/?id=2.

5. Cross, R.G., Higbie, J.A. and Cross, D.Q. (2009). Revenue management's renaissance: a rebirth of the art and science of profitable revenue generation. Cornell Hospitality Quarterly, 50, 56-81.

6. Duff, M. (2009). Consumers price sensitivity grows but with an everyday focus. CBS Interactive, Inc. Retrieved on March 1, 2010, from the BNET Industries website: http://industry. bnet. com/retail/10001288/consumers-price-sensitivity-grows-butwith-an-everyday-focus/.

7. Enz, C.A., Canina, L. & Lomanno, M. (2009). Competitive hotel pricing in uncertain times. CHR Reports, 9 (10), 4-14.

8. Enz, C.A., Canina, L. & Lomanno, M. (2004). Why discounting doesn't work: the dynamics of rising occupancy and falling revenue among competitors. CHR Reports, 4 (7), 5-25.

9. Feiertag, H. (1993). Quick discounting may lower perception of value. Hotel & Motel Management, 208 (19), 16.

10. Four Seasons Hotel (2010). Four Seasons Hotel Las Vegas. Retrieved on March 2, 2010, from http://www.fourseasons.com/lasvegas/.

11. Gu, Z. (1997). Proposing a room pricing model for optimizing profitability. International Journal of Hospitality Management, 16(3), 273-277.

12. Gu, Z. & Caneen, J.M. (1998). Quadratic models for yield management in hotel rooms operation. Progress in Tourism andHospitality Research, 4(3), 245-253.

13. Higley, J. (2003). Discounting isn't bad when it's done correctly. Hotel and Motel Management, 218(13), 8.

14. Kim, W.G., Han, J. & Hyun, K. (2004). Multi-Stage Synthetic Hotel Pricing. Journal of Hospitality and Tourism Research, 28(2), 166-185.

15. Kimes, S. (2009). Hotel revenue management in an economic downturn: results of an international study. CHR Reports, 9(12), 4-17.

16. Kotler, P., Bowen, J.T. & Makens, J.C. (2010). Marketing for hospitality and tourism (5th ed.). Upper Saddle River, NJ: Prentice Hall.

17. Lansner, J. (2010, January 8). Calif. hotel foreclosures quadruple. The Orange County Register. Retrieved January 28, 2010, from http://www.ocregister.com.

18. Lewis, S. (2009). Las Vegas: recessionary market trends in spotlight. Hotel & Motel Management, 224 (11), 8.

19. Lloyd-Jones, A.R. (2009). Stressed and distressed hotels: a review of the factors that can make a hotel more vulnerable to the market conditions. The Real Estate Finance Journal, 25(2), 8-11

20. Lomanno, M. (2008). Discounting rates leads to decreased product value. Hotel and Motel Management, 223(21), 22.

21. Mangan, E. & Collins, A. (2002). Threats to brand integrity in the hospitality sector: evidence from a tourist brand. International Journal of Contemporary Hospitality Management, 14(6), 286-293.

22. Marriott International, Inc. (2010). Grand Hotel Marriott Resort, Golf Club & Spa. Retrieved on March 2, 2010, from http://www.marriott.com/hotels/travel/ptlalgrand-hotel-marriott-resort-golf-club-and-spa/.

23. Martinez, C. (2009). Rate integrity is critical in 2009: hotels should focus as much on the integrity as the fee. New England Hotel Magazine, 19-20.

24. McCann, J. (2000). Revenue management is as hot as it gets for Hilton employee. Hotel and Motel Management, 215(19), 140.

25. Nassauer, S. (2010, February 4). Less housekeeping, more perks. The Wall Street Journal. Retrieved March 1, 2010, from http://online.wsj.com/article/SB10001424052748704259304575043160856596220.html?mod=WSJ_Hospitality_leftHeadlines.

26. Nevistas and Raff, H. (2009). Five alternatives to discounting prices. Hotel News Resource. Retrieved January 26, 2010, from http://www.hotelnewsresource.com/article39931.html.

27. Nevistas and Salerno, N. (2009). Mastering the art of hotel revenue management part one..the mission. Hotel News Resource. Retrieved January 11, 2010, from http://www.hotelnewsresource.com/article40796.html.

28. Office of Travel and Tourism Industries (n.d.). The state of U.S. travel and tourism industries. Retrieved January 31, 2010, from http://tinet.ita.doc.gov/pdf/state_of_travel_tourism_YTD.pdf.

29. Orlando/Orange County Convention & Vistors Bureau, Inc. (2009). State of the market. Retrieved December 9, 2009, from http://www.orlandoinfo.com/research/market/index.cfm.

30. Pan, C-M. (2007). Market demand variations, room capacity and optimal hotel room rates. International Journal of Hospitality Management, 26, 748-753.

31. Reed Business Information (2010a). Hotels in survival mode: guests can't help but notice 400,000 lay-offs, upgrades on hold. Hotels: The Magazine of the Worldwide Hotel Industry. Retrieved on March 20, 2010, from http://www.hotelsmag.com/lexisnexis/9881073 - Hotels_in_survival_mode_Guests_can_t_help_but_notice_400_000_layoffs_upgr ades_on_hold.php.

32. Reed Business Information (2010b). U.S. RevPAR fell 17% in 2009, STR reports. Hotels: The Magazine of the Worldwide Hotel Industry. Retrieved on January 31, 2010, from http://www.hotelsmag.com/article/446420- U_S_RevPAR_Fell_17_In_2009_STR_Reports.php?rssid=20620&q=2009+RevP AR .

33. Relihan, W.J., III (1989). The yield-management approach to hotel-room pricing. Cornell Hotel and Restaurant Administration Quarterly, 30(40), 40-45.

34. Revenueyourhotel.com and HSMAI Revenue Management Advisory Board (n.d.). Top 10 revenue management opportunities for 2010. Retrieved February 23, 2010, from http://www.revenueyourhotel.com/Revenue-Management/10.html.

35. Russell, K. (2009). The recession's effects on seattle's hotel industry. Retrieved on January 31, 2010, from http://www.4hoteliers.com/4hots_aprt.php?mwl=4372&tfn=F. Schmidgall, R.S. (2006). Hospitality industry managerial accounting (6th ed.). Lansing, Michigan: Educational Institute of the American Hotel & Lodging Association.

36. Shedlock, M. (2009). Wages and salaries fell 4.7%, most on record. Retrieved on February 3, 2010, from http://globaleconomicanalysis.blogspot.com/2009/08/wages-and-salaries-fell-47- most-on.html.

37. Smith, R. (2009). Pricing power evaporates: The U.S. lodging industry demonstrated signs of resisting discounting during the early part of the recession, but the past six months have been a different story. Cornell Hospitality Quarterly, 50 (3), 286- 288.

38. Steed, E. & Gu, Z. (2005). An examination of hotel room pricing methods: practiced and proposed. Journal of Revenue and Pricing Management, 3(4), 369-379.

39. Taylor, W.J. & Kimes, S. E (2010). How hotel guests perceive the fairness of differential room pricing. CHR Reports, 10 (2), 6-14.

40. Twelve Vacation Days Publishing (2010). 20 recession travel rules – surefire ways to get more vacation for less. Retrieved on February 1, 2010, from http://www.wejustgotback.com/ default.aspx?mod=recessiontravelrules_1.

41. U.S. Bureau of Labor Statistics, (n.d.). Employment situation summary. Retrieved on February 3, 2010, from http://www.bls.gov/news.release/empsit.nr0.htm.

42. Vinod, B. (2004). Unlocking the value of revenue management in the hotel industry. *Journal of Revenue Pricing and Management*, 3(2), 178-190.

43. Waller, F. (1997). The 'three c's ' of pricing. *Hotels*, 31 (9), 30.

44. Wikipedia (2010). Integrity. Retrieved February 28, 2010, from http://en.wikipedia.org/wiki/Integrity.

45. Withiam, G. (2001). A "4-c" strategy for yield management. *CHR Reports*, 4-16.

46. Wolfe,K. & Boger, C., Jr. (2001). Investigating rate integrity in convention cities: comparing central reservation and internet rates. *Journal of Convention and Exhibition Management*, 3(2), 15-31.

INDEX